Individual Latent Error Detection (I-LED)

Making Systems Safer

T0314437

Transportation Human Factors: Aerospace, Aviation, Maritime, Rail, and Road Series

Series Editor: Professor Neville A. Stanton, University of Southampton, UK

Automobile Automation
Distributed Cognition on the Road
Victoria A. Banks and Neville A. Stanton

Eco-Driving
From Strategies to Interfaces
Rich C. McIlroy and Neville A. Stanton

Driver Reactions to Automated Vehicles
A Practical Guide for Design and Evaluation
Alexander Eriksson and Neville A. Stanton

Systems Thinking in Practice
Applications of the Event Analysis of Systemic Teamwork Method
Paul Salmon, Neville A. Stanton and Guy Walker

Individual Latent Error Detection (I-LED)
Making Systems Safer
Justin R.E. Saward and Neville A. Stanton

For more information about this series, please visit: https://www.crcpress.com/Transportation-Human-Factors/book-series/CRCTRNHUMFACAER

Individual Latent Error Detection (I-LED)

Making Systems Safer

Justin R.E. Saward
Neville A. Stanton

CRC Press
Taylor & Francis Group
Boca Raton London New York

CRC Press is an imprint of the
Taylor & Francis Group, an **informa** business

CRC Press
Taylor & Francis Group
6000 Broken Sound Parkway NW, Suite 300
Boca Raton, FL 33487-2742

First issued in paperback 2023

© 2019 by Taylor & Francis Group, LLC
CRC Press is an imprint of Taylor & Francis Group, an Informa business

No claim to original U.S. Government works

ISBN 13: 978-1-03-257032-7 (pbk)
ISBN 13: 978-1-138-48279-1 (hbk)

DOI: 10.1201/9781351056700

Library of Congress Cataloging-in-Publication Data

Names: Saward, Justin R. E., author. | Stanton, Neville A. (Neville Anthony), 1960- author.
Title: Individual latent error detection (I-LED) : making systems safer / Justin R.E. Saward and Neville A. Stanton.
Description: Boca Raton, FL : CRC Press/Taylor & Francis Group, 2018. | Series: Transportation human factors : aerospace, aviation, maritime, rail, and road series | Includes bibliographical references and index.
Identifiers: LCCN 2018023903| ISBN 9781138482791 (hardback : acid-free paper) | ISBN 9781351056700 (e-book)
Subjects: LCSH: Airplanes--Maintenance and repair--Quality control. | Aeronautics--Human factors. | Errors--Prevention. | Fallibility. | Latent variables.
Classification: LCC TL671.9 .S48 2018 | DDC 629.134/6--dc23
LC record available at https://lccn.loc.gov/2018023903

Publisher's Note
The publisher has gone to great lengths to ensure the quality of this reprint but points out that some imperfections in the original copies may be apparent.

Visit the Taylor & Francis Web site at
http://www.taylorandfrancis.com

and the CRC Press Web site at
http://www.crcpress.com

For Victoria, Alexander and Clara; Dad, and in memory of Mum

Justin R.E. Saward

For Maggie, Josh and Jemima

Neville A. Stanton

Contents

Preface

The motivation for the research presented in this book came from the observation that aircraft maintenance engineers seemed to spontaneously recall errors that they had made. This recall happened in all sorts of places, such as the aircraft hangar, crew rooms, on the drive home as well as at home. When the recall happened, the aircraft engineers would either go and check the aircraft themselves or phone in to the hangar and report the error. Often as not, their recall would be correct and the remedial activities to correct the error would be undertaken. The authors of this book realised that this phenomenon is experienced by everyone all of the time. It is the 'did I lock my front door?', 'did I leave the cooker on?', 'did I lock my car?' kind of questions that seem to spontaneously pop into people's minds. This recall also happens for work-related activities, which has been called Individual Latent Error Detection (I-LED) in this book. The authors realised that the I-LED phenomenon had the potential to be incorporated into a safety management system, and required further study to understand how it could be used to make systems safer.

System-induced human error is one of the most significant factors in aircraft accidents. Human error is an inevitable by-product of performance variability caused by system failures. Undetected human error becomes a latent error condition that can impact system safety and therefore contribute to a future undesired outcome such as harm to people, damage to equipment or loss of output. I-LED refers to the detection of past errors by individuals who suffered the error via the seemingly spontaneous recall of past activity. An extensive review of literature covering human performance and aircraft maintenance error found that the phenomenon was as yet unexplored and therefore presents a new field for safety research and practice. The nature and extent of I-LED events is explored in this book in aircraft maintenance. Multi-process theory is developed and combined with a systems perspective to provide a theoretical framework upon which to conduct real-world observations of I-LED events in cohorts of naval aircraft engineers. The data from these real-world observations indicate time, location and other system cues act as I-LED trigger events. A time window of two hours seems to be critical for effective recall. Several practical interventions were designed and tested, with the 'Stop, Look and Listen' approach being most effective. The benefits of integrating I-LED interventions as an additional safety control within an organisation's safety system are presented. The research presented in this book offers a step-change in safety thinking by offering a new level of resilience within the workplace that has not previously been accounted for in organisational safety strategies. The applicability of I-LED to other aircraft maintenance environments, as well as wider applications, is considered in the conclusions.

We imagine that this book will be of interest to both researchers and practitioners. For researchers, we have opened up a new field of enquiry in 'human error'. This is an exciting endeavour, to be at the cutting edge of research. There is tension between systems theory and 'human error' that we have sought to resolve. Theorists will most likely find Chapters 2 and 3 most interesting. Empiricists and practitioners are most likely to find Chapters 4 to 7 of most use to them, where we elaborate on

I-LED and test the interventions. Chapters 8 and 9 are most likely to be of use to the safety community, who will want to implement I-LED in their own organisations. Unusually for research of this type, we actually perform a cost-benefit analysis of the interventions. Finally, we imagine that most readers will be interested in Chapter 10, where future applications of I-LED are considered. This book should be of interest and use to any safety critical organisation seeking to make their system safer.

Authors

Dr. Justin R.E. Saward is a chartered engineer and commander in the Royal Navy where he serves as an air engineer officer (AEO) and safety specialist. He holds an MSc in Human Factors & Safety Assessment in Aeronautics from Cranfield University and a safety-focused PhD from the University of Southampton. He was awarded the Honourable Company of Air Pilots (Saul) Prize for Aviation Safety Research in 2015, which is presented to an early career researcher who has advanced aviation human factors. He is currently head of safety strategy and a senior advisor on operational human factors for the Royal Navy, for which he designs safety transformation programmes and is also developing advanced safety management techniques for high-risk naval operations, encompassing air, land and sea domains.

Professor Neville A. Stanton, PhD, DSc, is a chartered psychologist, chartered ergonomist and chartered engineer. He holds the Chair in Human Factors Engineering in the Faculty of Engineering and the Environment at the University of Southampton in the United Kingdom. He has degrees in Occupational Psychology, Applied Psychology and Human Factors Engineering and has worked at the Universities of Aston, Brunel, Cornell and MIT. His research interests include modelling, predicting, analysing and evaluating human performance in systems, as well as designing the interfaces and interaction between humans and technology. Professor Stanton has worked on the design of automobiles, aircraft, ships and control rooms over the past 30 years on a variety of automation projects. He has published 40 books and over 300 journal papers on ergonomics and human factors. In 1998, he was presented with the Institution of Electrical Engineers Divisional Premium Award for research into system safety. The Institute of Ergonomics and Human Factors in the United Kingdom awarded him the Otto Edholm Medal in 2001, the President's Medal in 2008 and 2018, as well as the Sir Frederic Bartlett Medal in 2012 for his contributions to basic and applied ergonomics research. The Royal Aeronautical Society awarded him and his colleagues the Hodgson Prize in 2006 for research on design-induced flight deck error, published in The Aeronautical Journal. The University of Southampton has awarded him a Doctor of Science in 2014 for his sustained contribution to the development and validation of human factors methods.

List of Abbreviations

AAIB   Air Accidents Investigation Branch
AEO   Air engineer officer
AET   Air engineer technician
AFS   After flight servicing
ASIMS   Air safety information management system
AMCO   Air maintenance coordination office
BU   Bottom-up
C   Control
CBA   Cost benefit analysis
CFQ   Cognitive failures questionnaire
CIT   Critical incident technique
CoI   Cost of integration
DASOR   Defence air safety occurrence report
EGR   Engaged ground run
FAA   Federal Aviation Administration
GEMS   Generic error modelling system
GB   Gearbox
H   Hazard
HEI   Human error identification
HFACS   Human factors analysis and classification system
HFE   Human factors and ergonomics
HFI   Human factors integration
HSE   Health and safety executive
I-LED   Individual latent error detection
IR   Intentional review
JSP   Joint service publication
LE   Latent error
LED   Latent error detection
LES   Latent error searching
MAP   Military airworthiness publication
MEDA   Maintenance error/event decision aid
MF700   Aircraft document
MoD   Ministry of Defence
MRGB   Main rotor gearbox
MTF   Maintenance test flight
MxMHr   Maintenance man-hour
PCM   Perceptual cycle model
PIS   Participant information sheet
PM   Prospective memory
PPE   Personal protection equipment
PSF   Performance shaping factor
PTF   Partial test flight

ROI	Return on investment
RN	Royal Navy
RtL	Risk to life
SA	Situational awareness
SAS	Supervisory attentional system
SF	System failure
SLL	Stop, look and listen
SRK	Skill, rule or knowledge
STS	Sociotechnical system
TD	Top-down
TSM	Total safety management
UK	United Kingdom
UR	Unintentional review
VPF	Valuation of prevented fatality

1 Introduction

1.1 BACKGROUND

System-induced human error is recognised widely as the most significant factor in aircraft accidents, for which error is both inevitable and a frequent occurrence (Reason, 1990; Hollnagel, 1993; Maurino et al., 1995; Perrow, 1999; Wiegmann & Shappell, 2003; Flin et al., 2008; Woods et al., 2010; Amalberti, 2013). The potential consequences of system-induced error in safety critical contexts is universally understood, both in civilian and military environments where undetected human error leads to a latent error condition that can contribute to a future system failure (Helmreich, 2000; Shorrock & Kirwan, 2002; Wiegmann & Shappell, 2003; Flin et al., 2008; Reason, 2008; Woods et al., 2010; Aini & Fakhru'l-Razi, 2013). This can occur when there is inadequate control of human performance variability due to deficiencies in the safety system caused by sociotechnical factors that influence safety behaviour (Leveson, 2004; Woods et al., 2010). Rarely is an organisational accident the result of a single cause (Perrow, 1999; Amalberti, 2013); thus, it is often the confluence of more than one past error that can create a causal path to an organisational accident (Reason, 1997; Perrow, 1999; Matsika et al., 2013).

The detection of past errors has been observed amongst air engineers within UK naval aircraft maintenance, which does not appear to be wholly attributable to established safety management systems designed to defend against system deficiencies that can lead to human error. During the normal course of his employment as a serving Royal Navy Air Engineer Officer (AEO), the lead author observed occasions of workplace past errors that were detected at some point post-task completion through the recollection of past activity by the individual who suffered the error. This Individual Latent Error Detection (I-LED) phenomenon appeared to be spontaneous since recall was unplanned and occurred without making a conscious decision to review past activity or cued from a formal process check or through an independent inspection by a third party. Examples include the later realisation that a tool was not removed from the aircraft engine bay, an oil filler cap was not replaced after replenishing the reservoir, or the aircraft documentation was not completed correctly. More general everyday failures might be the spontaneous realisation that the gas hob had been left on or the door of their car or house was not locked. In all these examples, if left undetected, the latent error condition could contribute to unwanted consequences.

To understand the nature and extent of this phenomenon, an extensive literature review of wide-ranging publications shows system causes of human error effects have been researched widely, as has error avoidance and proximal detection, but no specific research or safety strategies appear to be available on I-LED. This confirms the phenomenon to be a novel concept, which is explored in this book.

System thinking transfers the emphasis from seeking individual human failings to understanding the network of sociotechnical factors that can cause system failures (Leveson, 2011). The systems view favours this macro-ergonomic approach to safety rather than the micro-ergonomic lens that can overly focus on individual human failings (Zink et al., 2002; Murphy et al., 2014). Thus, macro-ergonomic analysis explores interactions across sociotechnical systems or networks between elements comprising humans, society, the environment and technical aspects of the system such as machines, technology and processes (Emery & Trist, 1960; Reason & Hobbs, 2003; Woo & Vincente, 2003; Walker et al., 2008; Amalberti, 2013; Wilson, 2014; Niskanen et al., 2016). These networks can be complex in terms of the number interactions between systemic factors such as tools, equipment, procedures, decision-making, operator training and experience, and operating contexts (Edwards, 1972; Reason, 1990) and is where progressive safety strategies recognise that every element of the Sociotechnical System (STS) contributes to the organisation's safety goals through specific roles, responsibilities, relationships and safety behaviours (Flin et al., 2008; Woods et al., 2010; Plant & Stanton, 2016). Naval aircraft maintenance is typical of a complex sociotechnical network in a safety critical organisation, and thus macro-ergonomic analysis of safety factors impacting the human performance of naval aircraft engineers aligns to systems thinking, but a human-centred approach is essential to unlock knowledge on individual safety behaviour borne from I-LED events.

In the absence of specific research on the spontaneous recall of past errors, a new multi-process theoretical framework is developed from existing theories on Prospective Memory (PM), Supervisory Attentional System (SAS) and schema theory. Schema theory is characterised through the Perceptual Cycle Model (PCM), which describes a cyclic relationship between schema selection and sensory cues in the external world that trigger human actions (Neisser, 1976). The PCM therefore helps with understanding cognitive safety behaviours in the natural workplace environment (Smith & Hancock, 1995; Stanton et al., 2009a; Plant & Stanton, 2013a). The multi-process framework is combined with the systems perspective to observe the I-LED experiences of cohorts of naval aircraft engineers during normal aircraft maintenance activity, which is argued to offer a methodology that is congruent with progressive systems research in Human Factors and Ergonomics (HFE: Carayon, 2006). Using this methodology, naturalistic studies identify sociotechnical factors that facilitate I-LED events. The findings from these studies provide direction for the design and testing of several practicable I-LED interventions, which deliberately help aircraft engineers engage with system cues such as workplace objects and written words.

Organisational accidents are the consequence of system hazards that can cause harm to people, equipment and the environment when not controlled adequately due to deficiencies in safety controls designed to mitigate for the inevitability of performance variability (Reason, 1997; Leveson, 2011). A resilient safety system is dependent on adequate safety controls (Reason, 2008; Woods et al., 2010; Amalberti, 2013; Hollnagel, 2014). I-LED interventions described in this book act as an additional safety control mechanism to support safety resilience in an organisation. Here it is argued a total safety approach using I-LED interventions integrated within the overall safety system benefits resilience by providing further mitigation for human performance variability in the workplace. In support of this argument, new models for organisational resilience and operator competence are proposed to demonstrate how enhanced safety behaviours through I-LED interventions can help optimise safety in the workplace. To argue the costs versus benefits of integrating I-LED interventions within an existing safety system, financial costs and other resourcing implications are discussed. This completes the review of the nature and extent of the proposed I-LED phenomenon and its contribution to making the aircraft maintenance system safer through enhanced resilience.

1.2 AIM AND OBJECTIVES

The aim of this book is not to explain why errors occur or offer new error avoidance strategies; it is to understand the I-LED phenomenon from a systems perspective so that safety-related interventions can be developed that help optimise organisational safety resilience, thereby making the safety system safer. I-LED is the effect (dependent variable) to be observed with sociotechnical cues, present in the world, providing the triggers for recall. To achieve this aim, safety research presented in the following chapters is framed around five objectives:

- *Objective 1:* Using a human-centred systems approach, develop a theoretical framework to observe the I-LED phenomenon.
- *Objective 2:* Apply the theoretical framework to understand the nature and extent of I-LED events experienced by aircraft engineers working in their everyday maintenance environment.
- *Objective 3:* Identify practicable interventions that enhance I-LED events in safety critical contexts.
- *Objective 4:* Understand the effectiveness of I-LED interventions in the workplace.
- *Objective 5:* Assess the benefit of integrating I-LED interventions within organisational safety strategies to enhance resilience.

1.3 STRUCTURE OF BOOK

This book comprises ten chapters to address the five objectives using three linked observational studies to investigate the I-LED phenomenon in cohorts of naval aircraft engineers working in their normal operating environment.

Chapter 1: Introduction. This chapter introduces the I-LED phenomenon observed in naval aircraft maintenance before stating the aim and objectives followed by a summary of each chapter and a description of how I-LED can contribute to making the safety system safer through enhanced resilience.

Chapter 2: Application of Multi-Process Theory to I-LED Research. This chapter presents a review of human error detection literature and provides the context for research. Despite an extensive literature review, the nature and extent of this phenomenon is not understood fully and appears to be an under-explored safety field, whilst the causes of error and proximal error detection have been researched widely. To explore this phenomenon, a new theoretical framework is introduced based on a multi-process systems approach that combines theories on PM, SAS and schemata. Several examples from a UK military safety database are then analysed for existence of the phenomenon and evidence of the applicability of the multi-process approach. Thus, the intent is not to explain why human error occurs; it is to develop a theoretical framework upon which to observe how an individual who suffered an error later detects their error without any apparent deliberate attempt to recall past activity.

Chapter 3: Rationalising Systems Thinking with the Term 'Human Error' for Progressive Safety Research. During the review of human error detection in Chapter 2, it is noted that some concerns exist over the use of the term 'human error' in contemporary systems thinking, as it can be misused at the micro-ergonomic level to blame individuals rather than signpost the opportunity to tackle deficiencies within the STS that caused the error. This chapter looks to authorise the use of the term by considering literature on the subject. The term 'human error' is argued to remain meaningful when conducting HFE research from a systems perspective and therefore does not need to be excluded from progressive safety research such as I-LED. As with any lexicon, terms must be used correctly, for which human error is simply a subset of macro-ergonomic systemic factors that flags the requirement for HFE analysis of systemic factors. As such, the term is used universally in the current I-LED research presented in this book where aircraft engineers create safety through their own I-LED events, triggered by system cues available in the surrounding sociotechnical environment.

Chapter 4: Observing I-LED Events in the Workplace. To understand how to explore the nature and extent of I-LED in samples of naval aircraft engineers, the application of appropriate theory and observation methods is reviewed. To help ensure quality data are captured, and to remain flexible to emergent findings, a series of linked studies using a mixture of methodologies are introduced. This strategy also accommodates changes to the research design if emergent findings materialise that require a different methodology or revision to the current research. This chapter also recognises that conducting real-world observations in the workplace is challenging when compared to observations made in a laboratory setting, where strict experimental control can be achieved. However, advances in safety research come from exploring normative behaviours during normal operations, and this advantage is argued to outweigh the challenges of real-world research.

Chapter 5: A Time and a Place for the Recall of Past Errors. This chapter describes an exploratory study involving a target population of naval aircraft engineers using a questionnaire that was administered during group sessions. It was hypothesised that time, location and systems cues influence I-LED amongst engineers who had experienced the phenomenon. The systems view of human error is combined with a multi-process theory to explore the I-LED phenomenon, for which the findings suggest that the effective cognition of cues distributed across the STS triggers post-task I-LED events. The exploratory study makes the link to safety resilience using a systems approach to minimise the consequences arising from latent error conditions and provides direction for further research.

Chapter 6: A Golden Two Hours for I-LED. The findings from Chapter 5 provide direction for a second study described in this chapter. The detection of errors post-task completion is observed in a further cohort of naval aircraft engineers using a diary to record work-related I-LED events. The systems approach to error research is again combined with multi-process theory to explore sociotechnical factors associated with the I-LED phenomenon. Perception of cues in different environments facilitates successful I-LED where the deliberate review of past tasks within a time window of two hours of the error occurring, and whilst remaining in the same sociotechnical environment to that which the error occurred, appears most effective at triggering recall. Several practicable interventions offer potential mitigation for latent error conditions, particularly in simple everyday habitual tasks. The chapter concludes with the view that safety critical organisations should look to design further resilience through the application of I-LED interventions that deliberately engage with system cues across the entire sociotechnical environment to trigger recall, rather than relying on consistent human performance or chance detections.

Chapter 7: I-LED Interventions: Pictures, Words and a Stop, Look, Listen. Human error is a by-product of performance variability caused by system failures, for which the accumulation latent error conditions can contribute to an undesired outcome. Knowledge of the nature and extent of the I-LED phenomenon gained from the two earlier studies are combined to design and test several practicable I-LED interventions aimed at defending against the networking of latent errors that can lead to an organisational accident or degradation in system performance. The I-LED interventions are tested on a new cohort of naval aircraft engineers. Of those tested, a simple stop, look and listen intervention is found to be the most effective at making the system safer. The application of I-LED interventions is argued to offer a further step-change in safety thinking by helping to manage system-induced human error effects through the deliberate engagement with cues that trigger recall, thereby facilitating Safety II events. Successful I-LED limits occasions for adverse outcomes to occur, despite the presence of existing control strategies, and thus should be of benefit to any safety critical organisation seeking to enhance resilience in their existing safety system.

Chapter 8: A Total Safety Management Approach to System Safety. To understand how I-LED interventions might benefit an organisation's safety system, literature on human-centred safety strategies is reviewed. It is argued a truly resilient safety system comes from a total safety approach that optimises sociotechnical controls at the organisational level through to the management of safety by individual operators in the workplace. A review of literature finds few models that describe an agreed safety strategy for managing organisational resilience. Thus, a Total Safety Management (TSM) model is proposed that highlights the hierarchical relationship between safety 'as done' by competent operators through to the 'as designed' safety system. It is further argued resilience is dependent upon the system's ability to promote safe behaviours during interactions between humans and operating environments, which is predicated on the presence of competent operators in the safety network.

Chapter 9: Assessing the Benefits of I-LED Interventions. This chapter considers the costs versus benefits of integrating I-LED interventions within an organisation's existing safety management system. Cost Benefit Analysis (CBA) literature indicates that the main benefit of I-LED interventions comes from their integration with the organisation's safety system to form part of an enduring long-term safety strategy to engender and control safety behaviours in the workplace, which is discussed in the previous chapter. The integration of I-LED interventions as an additional safety control to help mitigate for system failures offers physical benefits in terms of reduced injuries or death, equipment damage and economics gains as well as intangible socio-political effects and non-technical attributes such as improved operator empowerment and job satisfaction. It is found that I-LED interventions are difficult to cost in financial terms due to often intangible cost data, although it is argued the typical interventions discussed in this chapter do not attract significant costs to introduce and maintain long term. Analysis of safety failures recorded in a military database show several occasions where the perceived safety severity of a reportable event was high or medium. Here it is argued that use of an I-LED intervention might have prevented the safety failure, which suggests significant Return on Investment (RoI) is achievable for safety critical organisations aiming to maximise the heroic abilities of its operators.

Chapter 10: Conclusions and Future Work. The aim of this book is to present safety research that contributes to making aircraft maintenance safer by understanding the nature and extent of the I-LED phenomenon and its benefit to safety resilience. The final chapter concludes the book with a summary of the safety research carried out against the five objectives, which support the aim. The novel contributions of this research are reviewed along with an evaluation of the approach to research and directions for future work. Concluding remarks offer a final statement of how I-LED interventions should be integrated within the safety system to make aircraft maintenance safer, as well as other safety dependent organisations such as energy suppliers, transportation and medical contexts.

1.4   MAKING AIRCRAFT MAINTENANCE SYSTEMS SAFER

This analysis in this book argues safety systems can be safer through exploitation of the I-LED phenomenon. This advances systems thinking and should benefit any safety critical organisation seeking gains in safety resilience. Multi-process theory combined with a macro-ergonomic perspective is argued to provide a suitable theoretical framework upon which to observe the I-LED experiences of cohorts of naval aircraft engineers working in their everyday environment. This approach confirms the presence of the I-LED phenomenon, for which I-LED events are likely to offer further mitigation for human-error effects resulting from deficiencies in safety strategies designed to control hazards in the workplace. The deliberate review of past activity within a time window of two hours of the error occurring and whilst remaining in the same sociotechnical environment to that which the error occurred appears most effective at recalling past errors. Time, location and other system cues facilitate these I-LED events. The detection of work-related latent error conditions is also found to occur when in non-work environments such as at home or driving a car;

indicating distributed cognition extends across multiple sociotechnical networks. I-LED is found to be common in simple everyday habitual tasks carried out alone where perhaps individual performance variability is most likely to pass unchecked, with the potential for errors to pass undetected. Application of a Cognitive Failures Questionnaire (CFQ) confirms naval aircraft engineers exhibit normal cognitive behaviours associated with wider populations of skilled operators. The testing of several I-LED interventions, designed to focus operator attention on system cues such as equipment or written words, shows a stop, look and listen intervention to be most effective at detecting past errors in the workplace. I-LED interventions can therefore act as an additional system control that helps defend against latent error conditions, which contributes to resilience within the safety system. The safety system needs to capture this new control strategy as part of integrated safety solution to control hazards where I-LED interventions can be included in a safety strategy managed locally by individual operators using system cues present in the world.

A new model depicting organisational safety resilience highlights the benefit of I-LED interventions if integrated within the overall safety system. An underpinning competence model also argues that there are four key components that need to be considered in the design of the safety system to help control performance variability seen in individual operators, who need to be suitably qualified, experienced and current or familiar with the task in-hand, possess personal readiness for work and be risk confident. A further review of literature assesses the costs versus benefits of introducing new safety interventions. Any safety intervention, including I-LED, should be founded in theory and form part of an enduring long-term safety strategy to engender and control safety behaviours in the workplace. However, it is problematical to calculate the benefit of I-LED in financial terms due to a lack of tangible cost data, although the I-LED interventions discussed are not thought to require significant financial or organisational costs to integrate within the safety system and maintain long term. Arguably, I-LED interventions tackle latent error conditions that lie hidden and propagate through the entire sociotechnical network and can, therefore, offer wider benefits in terms of reduced injuries or death, avoiding equipment damage and economics gains as well as socio-political effects and non-technical attributes such as improved operator empowerment. Analysis of UK military safety data highlights several safety occurrences where the perceived severity was high or medium. It is argued that use of I-LED interventions might have prevented the occurrence, and thus they are thought to offer significant return of investment (ROI) to safety critical organisations aiming to maximise the heroic abilities of its operators through I-LED interventions.

Overall, the findings from the safety research presented in this book provide new direction on how to make aircraft maintenance safer by advancing systems thinking through the exploration of the I-LED phenomenon. I-LED interventions derived from observations of naval aircraft engineers are likely to enhance overall safety resilience by offering further mitigation for undetected errors that can lie hidden the safety system. This should make the safety system safer. The I-LED ability of naval aircraft engineers is shown to be typical of skilled operators. Thus, the safety knowledge gained from research on this population should generalise to other aircraft maintenance organisations and wider safety dependent organisations employing similar cohorts of skilled workers.

2 Application of Multi-Process Theory to I-LED Research

2.1 INTRODUCTION

A human error detection phenomenon has been observed amongst naval aircraft engineers within UK operational helicopter squadrons, which does not appear to be wholly attributable to established safety mechanisms designed to defend against system failures leading to human error. The observations, made by the lead author who is a serving Royal Navy Air Engineer Officer (AEO), include examples such as the experienced air engineer who replenishes the aircraft engine oil during routine aircraft servicing but fails to replace the oil filler cap or who installs an aircraft component incorrectly during maintenance. Post the error event, and at some point later (minutes through to days) whilst resting in the crew room or continuing with other work say, the past error was often detected through some seemingly spontaneous recollection of past activity by the individual who suffered the error; upon which the individual was compelled to instigate a recovery. Studies have shown errors that pass undetected become latent error conditions where the impact of the error may not be immediately obvious due to delayed effects (Reason, 1990; Graeber & Marx, 1993; Lind, 2008), the effect being a causal path to a future unwanted safety outcome such as harm to people or equipment (Reason, 1997; Wiegmann & Shappell, 2003; Aini & Fakhru'l-Razi, 2013). The actions of the observed aircraft engineers resulted in the detection of individual latent error conditions before the system failure led to harm. The presence of the phenomenon, therefore, resulted in a safer system, and this provided the catalyst to investigate further.

After an extensive review of the literature, the nature and extent of the observed Individual Latent Error Detection (I-LED) phenomenon appears to be an under explored safety field and is therefore not understood fully. The literature review is summarised in Table 2.1, which included wide-ranging publications describing human error and recovery in cognitive and systems terms, as well as a review of the UK's Aviation Safety Information Management System (ASIMS) database. How individuals come to suffer error effects is well covered in the literature, where human error is widely reported to be inevitable and a daily occurrence (Norman, 1981; Reason, 1990; Hollnagel, 1993; Perrow, 1999; Wiegmann & Shappell, 2003; Flin et al., 2008; Woods et al., 2010). If human error is a common by-product of human performance, it is anticipated the proposed I-LED phenomenon is also prevalent.

The first objective in the current safety research is to understand the proposed I-LED phenomenon through the development of a theoretical framework using a

TABLE 2.1
Syntax Used in Literature Search

Syntax	Search Engine	Count	Action	Cited Articles
In title: individual AND	Scopus	18	All reviewed	0
latent AND error(s) AND	University of	20	All reviewed	
detection OR recovery	Southampton DelphiS			
	Google Scholar	6	All reviewed	
	Web of Knowledge	6	All reviewed	
In topic: human OR error	Scopus	848	Top 500 citations	69
detection OR recall NOT	DelphiS	220	All reviewed	
software NOT computer(s)	Google Scholar	1760	Top 500 citations	
NOT computing				
In topic: aviation OR aircraft	Scopus	395	All reviewed	15
AND maintenance AND	DelphiS	369	All reviewed	
error(s)	Google Scholar	596	All reviewed	

multi-process systems approach, which combines theories on schemata, prospective memory and attentional monitoring. The following chapters make no attempt to explain why human error occurs, as this is already covered widely in existing literature; it is to develop a theoretical framework upon which to explain how I-LED events occur without any apparent deliberate attempt to recall past activity.

2.2   CONTEXT FOR SAFETY RESEARCH

UK naval aviation is conducted in dynamic and complex environments, delivered around the globe from land bases in the United Kingdom with full aircraft support facilities to deployed temporary airfields with very limited facilities, and from large multi-aircraft carriers to small single-aircraft ships. Typical examples of naval aircraft maintenance environments are provided in Appendix A. The naval aircraft engineer operates within these operating contexts to deliver safe and effective aircraft maintenance and ground support to flying. Tasks vary significantly as the engineer transits between: the maintenance office where aircraft documentation is completed and tasks planned; the maintenance hangar; stores for parts; issue centre to collect tools; and the aircraft operating line (ramp) or ship's flight deck to launch, turn around and also service aircraft. This sees the aircraft engineer (not just within the military) perform a great number of disparate activities during the working day, impacted by: time pressures; extremes of weather; changing maintenance requirements due to emergent work or changes to the flying programme; and resource constraints in terms of equipment, spares and people (Latorella & Prabhu, 2000; Reason & Hobbs, 2003; Patankar & Taylor, 2004; Flin et al., 2008; Rashid et al., 2010; Woods et al., 2010). For naval aviation maintenance specifically, the context is very similar to that of commercial aviation yet compounded further due to operating aircraft from temporary airfields with very limited resources, operating from a moving platform whilst embarked in a warship, working on armed aircraft and significant operational imperatives. Away from the aircraft, the naval aircraft engineer will also be loaded further with extraneous duties, unrelated to aircraft maintenance, such as acting as force protection or assisting with ship general duties when deployed, all of which place additional demands on the engineer, which can lead to human performance variability that influences error rates. Additionally, the total effects of man–machine interactions with advanced technologies, common in military contexts, may not be known (Kontogiannis, 1999). Thus, resilience to all forms of human error is needed to avoid or mitigate for when error effects occur (Reason, 2008; Woods et al., 2010).

To construct a new theoretical framework to observe this phenomenon, the nature of human error is reviewed before describing a multi-process approach that combines theories on prospective memory, attentional monitoring and schemata with a recognised error categorisation format. Several examples from a UK military safety database are then used to facilitate initial exploration of theory as a precursor to further research via later real-world studies aimed at developing system interventions to mitigate for latent error conditions.

2.3   NATURE OF ERROR

2.3.1   HUMAN ERROR

Human error generally describes situations where either safety or the effectiveness of a task has been compromised due to human performance issues (Reason, 1990; Hollnagel, 1993; Amalberti, 2001; Wiegmann & Shappell, 2003). There are many definitions for human error. For example, Dhillon (2009, p. 4) defined error as 'the failure to perform a specified task (or performance of forbidden action) that could result in disruption of scheduled operations of damage to equipment and property'. UK defence aviation considers error to have occurred when 'an aircraft or system with human interaction fails to perform in the manner expected' (ASIMS, 2013b, p. 4), and in the maintenance context, Graeber and Marx (1993, p. 147) referred to 'an unexpected aircraft discrepancy (physical degradation or failure) attributed to the actions of the aircraft maintenance engineer'. Reason (1990, p. 9) defined error as 'a generic term to encompass all those occasions in which a planned sequence of mental or physical activities fails to achieve its intended outcome'. He also noted that people are not necessarily error prone, just more subject to error prone situations or conditions. Woods et al. (2010, p. 239) proposed error to be simply *the causal attribution of the psychology and sociology of an event*.

By citing just a few definitions, it appears defining human error is challenging, where ambiguity and a sense of hindsight are often present. Indeed, there has been some concern that the term human error can lead to blaming individuals for system fallings as opposed to tackling wider macroergonomic causes, which is where organisational safety strategies should seek knowledge to make the greatest safety gains (Zink et al., 2002; Dekker, 2014). This concern is addressed in Chapter 3, which argues the term human error remains meaningful in safety analysis from a systems perspective provided it is simply used as the signpost to investigate system failures that cause error. Dhillon's (2009) definition doesn't refer to error impacting the safety of people and seems to suggest error is exclusively damage or delays to scheduled operations. The UK Defence definition links error to some unspecified performance expectation, whilst Graeber and Marx (1993) suggest error to be specific to aircraft discrepancies only (rather than the totality of all aircraft maintenance activities). Hollnagel's (1993) analysis of human reliability in high-risk organisations also found human error difficult to define with most definitions having only little utility since they are often underspecified and contextually vague. In his later work on error, Reason (2008) agreed that there is no one universally agreed definition of error, whilst Woods et al. (2010) regard defining human error as remarkably complex with most error definitions being subjective. Thus, for current safety research, error is argued simply to be a colloquial expression used to flag error effects or erroneous actions that must be contextualised and explained against system causes (i.e., situational error) to be a meaningful.

2.3.2 IMPACT OF HUMAN ERROR IN AIRCRAFT MAINTENANCE

Human error effects are the most significant factor in aircraft accidents for both the military and civilian aviation organisations where errors occur regularly (Wiegmann & Shappell, 2003; Flin et al., 2008; Reason, 2008; Woods et al., 2010). System-induced human error is a normal by-product of human performance and a hazard to normal operating conditions that cannot be eradicated, deserving constant enquiry to identify and manage system failures causing error before an unwanted event permeates through the safety system (Reason, 1990, 2008; Amalberti & Barriquault, 1999; Helmreich et al., 1999; Dekker, 2006; Gilbert et al., 2007). Most estimates associate human error with 70%–90% of accidents (Hawkins, 1987; Hollnagel, 1993; Helmreich, 2000; Adams, 2006). UK military aviation data indicate at least 70% of naval aviation safety occurrences are attributable to human errors (ASIMS, 2013a). For aircraft maintenance specifically, analysis of a major airline showed the distribution of 122 maintenance errors over a period of 3 years to be omissions (56%), incorrect installations (30%) and wrong parts (8%) (Graeber & Marx, 1993). The Civil Aviation Authority (CAA, 2009) commissioned a study of UK civil aviation mandatory occurrence reports involving jet aircraft (years 1996–2006), which found incorrect maintenance actions and incomplete maintenance contributed 53.1% and 20.7%, respectively (total n = 3284 mandatory reports). Physical examples include: loose objects left in the aircraft; fuel caps unsecured; missing components or incorrect installation; and cowlings or access panels unsecured (Latorella & Prabhu, 2000). Notably, the latter example occurred to an Airbus A319-131 where the engine cowl doors were left in the unlatched condition (Figure 2.1) and resulted in the doors detaching in flight (AAIB, 2015).

In the case of incorrect installation, the classic example is the accident that occurred to a British Airways BAC 1-11 (Figure 2.2) that suffered an explosive decompression at cruising altitude (AAIB, 1990). Several of the attachment bolts securing the left-hand pilot's window were sized incorrectly during overnight maintenance on the aircraft. Once the aircraft was in flight and therefore pressurised at its cruising altitude, the window suffered a catastrophic failure and near loss of life occurred. Thus, the effects of human error are present in aircraft maintenance and potentially are a significant factor impacting the safety success and effectiveness of maintenance tasks if errors pass undetected.

FIGURE 2.1 Example engine cowl detached in flight. (Picture reproduced with permission from AAIB report 1/2015 involving Airbus A319-131, G-EUOE.)

FIGURE 2.2 BAC1-11 accident involving BA5390. (This image is copyright of Shutterstock/Rex.)

2.4 EXISTING ERROR DETECTION RESEARCH

Aircrew error has traditionally received much attention (Latorella & Prabhu, 2000), as the benefits are clear, although there is now global recognition that maintenance error also poses a serious hazard to safety that drives the need for system barriers or defences (Hobbs & Williamson, 2003; Reason & Hobbs, 2003). Here, research has moved to systemic factors as opposed to individual factors to avoid or mitigate for error since the human condition of error is inevitable, and it is seen as incumbent upon the organisation to 'protect' itself from human error effects (Dekker, 2006; Reason, 2008; Woods et al., 2010). However, it is argued that the systems perspective needs to maintain a human-centred approach to safety research. For example, understanding how individuals self-detect and recover from past error reveals a path to error protection in systemic contexts that can only come from knowledge of how individual safety behaviours integrate with the physical environment and overall safety system.

Woods (1984) studied operator performance during simulated emergency events for a nuclear power plant and found that two-thirds of errors across emergency scenarios went undetected, half the execution errors (slips and lapses) were detected by the operators and no mistakes (planning errors) were detected without intervention from an external agent. Error types are discussed in Section 2.5.2, where the link to schema theory is explored. Here, a slip refers to a task carried out in error (such as dropping a tool) and a lapse involves a task where the required action is omitted (such as forgetting to close a servicing panel). Mistakes occur when the action is carried out incorrectly due to flawed knowledge about the task or an incorrect rule is selected (Reason, 1990). Further, Kontogiannis and Malakis (2009) examined error detection rates in various aviation studies (Wioland & Amalberti, 1996; Doireau et al., 1997; Sarter & Alexander, 2000) and found that error detection rates appeared lower for mistakes (planning errors), yet higher for slips and lapses (execution errors). Sellen (1994) argued that slips are more often detected proximal to the error through the action itself and mistakes through an outcome that requires external intervention (through a procedure or third party).

A Maintenance Environment Survey Scale (MESS) was applied to Australian Army aircraft maintenance (Fogerty et al., 1999) and found 79% (n = 448) of respondents admitted to making errors that they self-detected and 50% to making errors that were detected by supervisors. Patel et al. (2011) studied error detection and recovery in the critical care domain using both experimental and real-world approaches. They showed that error detection and correction (recovery) are dependent on expertise and on the nature of the everyday task of the clinicians concerned, where most of their errors were detected and recovered whilst remaining at work. Similarly, a Wilkinson et al. (2011) study on error detection and recovery in dialysis nursing also found that expert nurses develop a special ability to detect and recover from their errors, and the nature of the error was dependent upon the nature of the task. An observational study of aircrew error by Thomas (2004) showed that around half of the errors went undetected by experienced crew, although few of these errors led to undesired aircraft state (unwanted outcome). Amalberti and Wioland (1997) showed errors made by highly trained operators (such as aircrew) can be frequent, yet most are either inconsequential or detected and corrected before leading to an undesired outcome. Blavier et al. (2005) also found the number of execution errors increases significantly with task complexity but so does their detection, whilst the number of planning errors and their detection are unaffected by complexity. These findings relate to proximal error detection, yet an operator can experience spontaneous belief that an error has occurred but be unsure of the nature of the error (Zapf & Reason, 1994) and search for information about the past activity. Here, Zapf and Reason considered that there are three ways errors are detected: conscious self-monitoring (planned detection); external environmental cue (unplanned); or a third party such as a supervisor when checking a safety critical task (again, planned detection). What is of interest when exploring the I-LED phenomenon is the seemingly spontaneous detection discussed earlier due to some unexplained unconscious self-monitoring (unplanned detection).

The literature review in Table 2.1 found no research on individual detection and recovery of latent error conditions, and thus it appears to be an under explored safety field. Causes of human error have been researched widely, as has error avoidance (Kontogiannis, 1999), for which formal error identification tools or Human Error Identification (HEI) techniques exist to predict the likelihood of error in safety critical organisations (Stanton & Baber, 1996). Such techniques calculate the likelihood of error with associated risk indices, but by design, these techniques are concerned largely with the design of safe systems and not the analysis of latent error conditions. Additionally, HEI techniques do not account for attentional mechanisms (self-monitoring) in observable behaviours or take adequate account of error contexts (Hollnagel, 1993; Stanton & Baber, 1996). For an error that has occurred, research has considered proximal error detection and found weaknesses in the effective detection of these errors, without which recovery is not possible (Blavier et al., 2005). Such research captures planned detection and correction of error as opposed to the unplanned latent detection of errors that 'come to mind' later (Sellen et al., 1996; Kontogiannis, 1999). Detection of execution errors is higher (albeit proximal to the error event) than planning errors, whilst the ability to self-detect errors appears to be linked to expertise (although the expert is not immune to error), and when experiencing spontaneous belief that something is in error, the individual will search for information about the past task (i.e., searching in memory for potential past errors). Since error detection appears to be influenced by multiple sociotechnical factors, the following section will argue for a multi-process approach to observe the phenomenon.

2.5 ROLE OF SCHEMA IN ERROR

2.5.1 SCHEMA THEORY

Bartlett (1932) introduced schema as information represented in memory about our external world of objects, events, motor actions and situations. Norman (1981, p. 3) defined this information as 'organised memory units' for knowledge and skills that are constructed from our past experiences, which we use to respond to (interact with) external sensory data. In this respect, schema theory is arguably aligned with Human Factors and Ergonomics (HFE) as a broad science that considers all aspects of the human interaction with the external environment (Carayon, 2006). Specifically, the application of schema theory has been influential and effective for military applications, including aviation (Plant & Stanton, 2013a) and thus provides theory for application to the naval aviation maintenance context.

What we encode and commit to memory is determined by pre-existing schema and is an iterative process, which develops residual memory structures or genotype schema based on our general understanding of the physical environment (Reason, 1990; Stanton et al., 2009b; Plant & Stanton, 2012). Schema consists of 'slots' of data

(Rumelhart & Norman, 1983) for fixed compulsory values or variable optional values, which are specified when planning intention (see Prospective Memory). Data slots are dependent on the interaction with external sensory information in the physical environment and are part of the process to select required schema. Where the sensory information is not available or not attended to by the conscious, the data slot will be represented by a default value (Cohen et al., 1986). For example, when refuelling an aircraft: fixed data are fuelling equipment and procedures; variable data are aircraft and amount of fuel; and default data could be use of the incorrect fuel as the aircraft engineer did not know the aircrew needed a different fuel type for the next flight (for military operations, there are several fuel options). A properly encoded schema is therefore essential for planned intent to be carried out correctly, and as will be discussed later, it will be predicted that the nature of the schema data slots influences error detection.

Schemata are also hierarchical and interrelated where only the highest-level schema (parent) needs conscious activation for subordinate (child) schema to trigger autonomously; child schema being needed for particular skills (i.e., motor response) or knowledge of particular properties of objects and location (Norman, 1981; Mandler, 1985; Cohen et al., 1986; Reason, 1990; Baddeley, 1997). For example, when the aircraft engineer conducts aircraft flight servicing the high-level schema for this activity is activated consciously with subordinate schemata to open and close panels, operate equipment (motor skills) and navigate around the aircraft triggering automatically (and in sequence). This is an important hierarchical feature of schema theory as it provides an explanation as to how finite capacity is made available to attend to other unrelated functions such as talking to a colleague whilst working or thinking about the next maintenance task. Execution of this activity requires the bottom-up (BU) processing of external data from sensory inputs with top-down (TD) prior knowledge (schemata) to project a response (Neisser, 1976; Cohen et al., 1986; Plant & Stanton, 2013a). This processing of the external environment with internal pre-existing schema to achieve the required response captures elements needed for situational awareness (SA) and is essential for the execution of tasks without error (Plant & Stanton, 2013a); situation awareness being 'the cognitive processes for building and maintaining awareness of a workplace situation' (Flin et al., 2008, p. 17). Here, reality needs to match the mental template (activated schemata) for correct SA (Rafferty et al., 2013), for which ongoing monitoring by the operator is needed.

Endsley and Robertson (2000) hypothesised that schemata are influential in the possession of correct SA, where the perception of external data and application of the correct schemata delivers correctly executed responses. For the previous example, schema theory indicates that the BU/TD processing of well-practiced tasks can lead to confusion as the memory trace for the real activity (fuel required by the aircrew) has been derived from a past experience (schema for the fuel type most commonly used). Amalgamation of a different reality to the genotype schema held in memory can lead to error (Cohen et al., 1986) and is also the view of Reason (1990), who considered cognitive under specification between external data and schemata can manifest as errors. Thus, the important link between error and schema theory needs to be made before constructing a theoretical framework upon which to explore the I-LED phenomenon, facilitating the opportunity for SA re-gains.

2.5.2　Linking Schema Theory to Error

Norman's (1981) research on execution errors described the use of incorrect schema applied to a planned task, which can lead to unintentional slip of action (error effect). It is argued that this finding can be applied to Reason's error types from which specific error behaviour or phenotype schemata can be observed in the physical environment to assess human performance 'in the moment' of activity (Stanton et al., 2009b; Plant & Stanton, 2012). Rasmussen (1982) categorised human performance behaviours as skill, rule or knowledge based (SRK). Skill-based performance is characterised by a routine, well-practiced and expected task that requires little cognition as the operator is largely pre-programmed (motor response) to carry out the task. Rule-based performance is typified by familiar situations (but not as routine as for skill based) where the task has been trained for such that a semi-conscious processing is needed to achieve the task. Knowledge-based performance relates to novel difficult situations that require greater cognition for analysis and planning to achieve a task, for which previous experience or training supports task success; characterising the experienced operator (Maurino et al., 1995; Reason, 1997). Here, it is argued Rasmussen's SRK-based performance behaviours can be linked directly to the nature of the activated schema (Stanton & Salmon, 2009).

Reason (1990) expanded on Norman's (1981) seminal work to produce the Generic Error Modelling System (GEMS) shown in Figure 2.3.

The GEMS categorises execution failures during routine situations as slips; a well-practiced skill-based task, requiring little cognition, is carried out incorrectly. Lapses are similarly well-practiced skill-based tasks, requiring little cognition, yet not carried out (omission). Erroneous actions (error), associated with conscious task planning, are categorised as mistakes where deficiencies in judgement or inferential processes cause the failing to formulate the correct plan based on flawed knowledge and/or incorrect comprehension of recognised rules (Reason, 1990; Sellen, 1994). There are also occasions of intended actions where individuals (or teams) elect to not follow a procedure or develop uncontrolled working practices that have evolved locally. There are multiple and complex reasons (often seeming rational to the individuals and/or

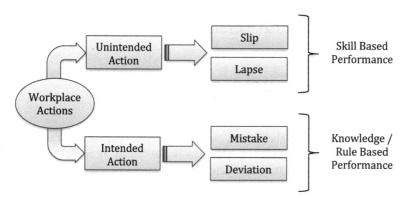

FIGURE 2.3　Generic Error Modelling System (GEMS). (Adapted from Reason, J., 1990. *Human Error*. Cambridge: Cambridge University Press.)

teams involved) for this type of action, for which the motivation is mostly personal or organisational gains. GEMS categorise this type of error as a violation or deviation, as shown in Figure 2.3, to better reflect the deviance from safe working practices, which become normalised over time (Vaughan, 1997). Reason (1990, p. 11) argued the GEMS categorisation of error helps focus attention on the interdependencies between 'local triggering factors and error behaviours' (phenotype), and of particular interest is behaviour leading to errors that pass undetected at the time they occur. Blavier et al. (2005) suggested that using Reason's GEMS has utility when describing error detection mechanisms as it can be related to attention and memory. For the current safety research, the link can be made to schema theory to develop a theoretical perspective for the observed phenomenon:

- Skill-based slips are a consequence of a correct schema(s) selected in memory but executed incorrectly (Norman, 1981; Rasmussen, 1982). One example would be the inadvertent operation of aircraft floatation bags during servicing;
- Skill-based lapses occur when the required schema(s) is not enacted (Rasmussen, 1982; Reason, 1990). One example would be forgetting to sign aircraft maintenance documentation;
- Knowledge- or rule-based mistakes occur due to a data slot(s) that is populated incorrectly, a wrong schema selected based on incorrect perception of external sensory data, or a correct high-level schema is selected, but the data slot(s) defaults incorrectly as external sensory information not available (Rasmussen, 1982; Rumelhart & Norman, 1983; Reason, 1990). One example would be carrying out a component replacement on the wrong aircraft. Additionally, a specific schema may not be available as the operator is faced with a novel situation. Here, the operator may look for a 'best fit' to achieve the task (Cohen et al., 1986) – for example, where a maintenance technique learnt for one aircraft type is used on a different type. Of note, this could be termed a deviation or violation.

The previous proposes an important link between error types (Reason, 1990) and schema theory, which can be applied to error detection. Additionally, storage of planned actions (selected schema) can be explained through prospective memory (see next section).

2.6   MULTI-PROCESS APPROACH TO I-LED

2.6.1   THE NEED FOR A MULTI-PROCESS APPROACH

Schema theory applies to the human responses needed to achieve a task (Norman, 1981), for which an attentional mechanism is needed to 'prepare' the schema(s) in memory to be enacted later and determine hierarchical interactions and trigger criteria. Here, it is argued that schema theory does not currently account for this activity, and thus, there must be other cognitive processes involved. The correct application of a particular schema at the correct time and to the correct task requires the preparation, activation, monitoring and feedback for task success. Multi-process theory provides a theoretical approach that may account for the total interaction with the external world where it is hypothesised that Prospective Memory (PM) is responsible for the preparation of schema (schema selection and trigger criteria) and a Supervisory Attentional System (SAS) for activation, monitoring and feedback (Norman & Shallice, 1986).

2.6.2 MULTI-PROCESS THEORY

A multiprocess approach, termed by Einstein and McDaniel (2005), argues task success is dependent on several mechanisms that include memory, monitoring mechanisms, task encoding, trigger cues and contextual factors (Brewer et al., 2010). Dismukes (2012) supported the multi-process concept but indicated that this does not necessarily account for everyday tasks that are habitual in nature and less likely to need a formal plan of action to be made such as for the aircraft engineer carrying out routine flight servicing. But this is a key feature of schema theory where well-practised execution tasks are largely automatic responses characterised by schemata for motor responses that may not require conscious control, although a level of monitoring is required (Green et al., 1996). Thus, task intent (prospective memory) and attentional mechanisms need to be considered, along with schema theory, to make multi-process predictions for the latent detection of situational error.

2.6.3 PROSPECTIVE MEMORY

Forming a plan of action enables intent to be stored in working memory until its execution (Palmer & McDonald, 2000; Dockree & Ellis, 2001). Originally recognised by Ebbinghaus (1885:1964) as a discrete type of memory, it is only in later years that working memory has been studied closely, from which recent literature has emerged that reports the concept of PM. This refers to occasions when intent has been formed and stored in working memory for later recall (Reason, 1990; Sellen, 1994; Ellis, 1996; Kvavilashvili & Ellis, 1996; Baddeley, 1997; Blavier et al., 2005; Dismukes, 2012), for which it is argued that schema selection, hierarchy and trigger criteria are the responsibility of PM. Reason (1990) reported that there could be a delay between intention forming in PM and taking action, which is particularly vulnerable to failure such as forgetting to carry out a planned action. Preparing cognitive information in memory from external cues in the world and then assessing task success requires some form of monitoring of the physical environment that is then needed to activate the schema as intended.

2.6.4 MONITORING THEORY

The Supervisory Attentional System (SAS) is responsible for matching of external sensory data (from the physical environment) with the selected schema, for which several schemata may be selected in a hierarchical pattern determined by the strength (importance) of the target activity (Plant & Stanton, 2013a). Studies on PM indicate that formed intentions (selected schema) are 'loaded' into memory to act upon later, which generates a 'to-do list' or internal marker (Sellen et al., 1996; Marsh et al., 1998;

Van den Berg et al., 2004). Dockree and Ellis (2001) found subsequent cancelling of the internal marker (deletion of a schema on the to-do list) may be regulated by the SAS described by Norman and Shallice (1986), which is influenced by abstract cues for thoughts and linguistics, sensory and perceptual cues, and psychological (emotional) state cues (Guynn et al., 1998; Bargh & Chartrand, 1999; Einstein & McDaniel, 2005; Mace et al., 2011).

Plant and Stanton (2013a) similarly argued that schema theory also predicts triggering in response to external sensory inputs (interaction with the physical environment) but the activation of a particular schema may also come from another schema; both of which are subject to interference from competing schemata that are similar or the selection of a new schema that has been deemed more important than the immediate activity. Since it has been argued that schemata operate largely autonomously, this provides an explanation for why some errors pass undetected, but it does not explain why such errors come back in mind later (for example, going upstairs to fetch a jumper but seeing the bed needs making, which you attend to before proceeding back downstairs without the jumper you planned to fetch). In the example of the oil filler cap not being replaced, it may be that a competing schema was considered more important, such as attending to leak that has been spotted on the aircraft. Here, the competing schema responding to the leak is seen as more important and enacted in preference to the schema for replacing the oil filler cap. Should this be true, it is not understood fully why the schema for the oil filler cap comes back to mind later such that the individual realises the missing cap. So, for this to occur, it is argued that some later review or inner feedback (Kontogiannis, 1999) of a selected schema(s) takes place that results in an involuntary memory, and this unplanned detection can be explained through SAS monitoring theory. Here, it is proposed that SAS monitoring leads to a review of completed tasks to confirm they have been executed correctly and, where necessary, updates established genotype schema and/or constructs new schema if the task was novel. This is a key feature of schema theory (Reason, 1990; Plant & Stanton, 2012) and could be seen as a 'housekeeping' function, which is essential for learning and acquiring experience. Further, it is argued that outcomes of this monitoring also explain the seemingly spontaneous recall of past errors (unplanned detection) as a consequence of post-task housekeeping or a conscious and deliberate search for potential latent errors where self-doubt comes to mind, and for which the latter is offered as Latent Error Searching (LES).

Bartlett's (1932) research found the accurate recall of past events was attributable to the ability to reconstruct schemata applied at the time a given task was executed. For the detection of slip errors, Norman (1981, p. 11) proposed a feedback mechanism exists 'with some monitoring function that compares what is expected with what has occurred' (arguably SAS monitoring). For successful error detection, this monitoring function detects any discrepancy between the selected schema and task execution for incorrect execution (genotype/phenotype schema mismatch) and/or genotype assimilation. Thus, it is this monitoring function that is of interest and thought to extend to lapses and mistakes and beyond the proximal to I-LED events, where past tasks comes to mind, upon which task success is questioned (Zapf & Reason, 1994; Sellen et al., 1996). Thus, SAS and schema theories combined may indicate that error detection is more successful when post-task housekeeping takes place in a physical environment that is the same as or similar to the physical environment in which the actual error occurred (or simply when there is a genotype/phenotype mismatch). Additionally, triggering conditions have been cited as critical for correct human performance (Norman, 1981) and are dependent upon the physical environment. Thus, the physical environment may also influence SAS monitoring for successful I-LED.

2.7 ERROR DETECTION AND RECOVERY EXAMPLES

2.7.1 ASIMS DATA

This chapter introduces a multi-process theoretical framework to explore I-LED events from the perspective of the aircraft engineer's interaction with the physical working environment. To facilitate initial exploration of the proposed theoretical framework prior to further research, ASIMS (2013a) was interrogated for examples of latent error conditions. This UK Ministry of Defence (MoD) restricted database contains approximately 30,000 air safety reports recorded using the Defence Air Safety Occurrence Report (DASOR) template (noting that ASIMS contains around a further 130,000 reports dating back to 1970 but prior to the introduction of DASORs in 1990). The template was constructed from two established HF templates in aviation: the Maintenance Error Decision Aid (MEDA) designed by Rankin and Allen (1996); and Human Factors Analysis and Classification System (HFACS) designed by Wiegmann and Shappell (2003). Each DASOR captures real-world error events and so offers a potentially rich secondary data source for re-analysis. Only Royal Navy (RN) aircraft maintenance reports were sampled to align with the chapter. This yielded 1933 reports that were sifted to just three, as the majority of reports did not cite latent error examples. This was not unexpected, since an individual may not see the need to report an error that has been recovered successfully, and is therefore potentially analogous to the pyramid heuristic (Heinrich, 1931) that suggested the vast majority of errors go unreported. Despite the scarcity of relevant DASORs, narratives from each report were individually analysed thematically before 'horizontal' comparison (Robson, 2011). This comparative technique (Cassell & Symon, 2004; Polit & Beck, 2004) highlighted common themes and optimised data extraction from the very limited number of reports (Schluter et al., 2008). Table 2.2 shows themes associated with latent error detection and recovery that were determined from the narratives of each of the three DASORs. Identifying information has been removed, and the bracketed text explains technical terms.

2.7.2 DASOR EXAMPLES

Subjectivity in reported narratives meant that it was difficult to determine themes with confidence, as textual data were underspecified. Additionally, to maintain the quality of deductive analysis, care was taken to ensure the researcher did not influence how textual data was interpreted, although this was also an advantage as the experience of the author was essential to exploit the maximum information from the technical

TABLE 2.2
Thematic Analysis of DASORs

DASOR Examples	Multi-Process Error Detection Themes			Error Recovery Themes
	PM Encoding	Trigger (Cueing) Condition	SAS Monitoring	
Example 1	'Flight Servicing'	'Lack of visual cues'	'Realised mistake'	'Reported it' 'Systems drained and replenished'
'During Flight Servicing, the AET [engineer] inadvertently selected Aeroshell 555 [oil] juniper rig [delivery system] to replenish the ECU [aircraft engine] oil system on all engines. Post replenishment, he realised his mistake and reported it immediately. The AET selected the correct equipment; unfortunately, he selected the rig with the wrong fluid. There is no evidence of distraction. He was not consciously fatigued, having returned home the night before. A contributory factor is the lack of visual cues to assist correct rig identification. Oil systems were drained and replenished.'				
Example 2	'Aircraft refuel'	'Obscured from view'	'Became apparent'	'Reported it' 'Items retrieved'
'AET conducted an aircraft refuel. On completion and once the fuel bowser had left the dispersal, it became apparent to the AET that he had left his night flying wands [torches] adjacent to the fuelling hose on the fuel bowser. AET recalls that when he placed the wands down that they were still illuminated. On completion of refuelling, the bowser driver pulled the cover down over the hose reel point, which would have obscured the wands that were inside it from view. As soon as the AET realised the items were missing, he reported it and items retrieved from the bowser.'				
Example 3	'Control cable change'	'Poor communication'	'Realised'	'Notified controller' 'Instigated an inspection'
'Following a tail rotor flying control cable change, the maintainer that carried out the task realised he had negated to remove the cable identification tags on completion. He immediately notified the watch controller who instigated an inspection, which revealed that the identity tags were still attached to the tail rotor cables. This incident was caused by poor communication between the supervisor and independent inspector.'				

lexicon used in the narratives (Schluter et al., 2008). For each, the importance of the text was questioned by interrogating each line with a 'so what' approach to allow themes to emerge (Robson, 2011 p. 479), although the extremely small number of reports means no statistical significance or strong evidence to assure the proposed theoretical framework can be claimed, since there is likely to be significant error in the findings and variance against the population. For each DASOR, the analysis served only to demonstrate the existence of the phenomenon and provide initial evidence that the multi-process theory proposed in this paper is applicable to the context being researched.

2.7.3 Link to Theoretical Framework

Themes highlighted in Table 2.2 were used to explore the applicability of the link between multi-process theory and the naval aircraft engineering context. Each DASOR describes well-practised maintenance tasks carried out by trained engineers, which included aircraft flight servicing, refuelling and changing a control cable. For these habitual tasks, it is offered that the PM element of the theoretical framework was established at task inception, and each reported error event was a subset (child schema) of the main maintenance activity (parent schema), and therefore, overall schema activation needed to be hierarchical and interrelated as each main task necessitated the sequencing and activation of sub tasks. In Example 1, it is argued the engineer tasked with carrying out flight servicing (parent schema) selected a replenishment rig containing the wrong oil, which had required activation of the child schema with particular knowledge of the equipment being used. The engineer tasked with an aircraft refuel in Example 2 activated a child schema that manifest as a motor response to place his torches on the fuel bowser during the refuelling sequence. For Example 3, the main task to replace tail rotor cables may have led to a latent error condition due to a failure to trigger the child schema with the knowledge that cable tags are to be removed upon completion of the cable replacement. The engineers who suffered these errors also detected the error later (post-task completion) whilst physically present in the hangar or outside on the line/dispersal (ramp); in each example, the engineer detected the latent error condition in the same or similar physical environment to that which the error occurred. Here, correct schema behaviour (phenotype schema) did not occur proximal to the maintenance task, as each engineer appears to have not detected important cues or sensory information in the physical environment such as the incorrect oil replenishment rig in the first example, location of the night flying torches in the second example, and communication between supervisor and independent inspector in the third example.

The themes 'realised mistake' and 'became apparent' may support the claim that the SAS monitoring element of the proposed theoretical framework is also present for the I-LED phenomenon to occur, where the schema mismatch (genotype/phenotype) is detected via a housekeeping function that occurs post-task completion (potentially giving rise to LES). Upon detection, each latent error condition was recovered successfully, which also minimised the consequences and avoided any potential 'bad' outcomes (Woods et al., 2010): the oil system was drained and replenished in Example 1; torches recovered in Example 2; and the tail rotor cable tags were removed in Example 3.

The limited DASOR data indicate the observed phenomenon has been experienced by aircraft engineers in the workplace, whilst the themes described at Table 2.2 suggest early evidence exists to link the proposed multi-process theoretical framework to the aircraft maintenance context. Within this framework, external cues (or triggers) across the sociotechnical system or network are argued to be of paramount importance, where both the nature of physical environment and the potential for vigilance decrement through missed cues (Reason & Hobbs, 2003) highlight the dependence on cue recognition for correct schema behaviour, which continues to be essential for successful I-LED.

2.8 SUMMARY

System-induced human error effects have been shown to be the most significant factor impacting the safety success of an organisation, yet the term human error is defined variously and is difficult to qualify or even measure objectively, and thus, a working definition for further research has been given. In reviewing literature to support the observed phenomenon, it is evident that much research has identified the vagaries of human error within an organisation, representing multiple occasions of system failures since error effects are inevitable and occur daily (Reason, 1990; Hollnagel, 1993; Norman, 1993; Maurino et al., 1995; Perrow, 1999; Wiegmann & Shappell, 2003; Woods et al., 2010). When there is a delay or absence in feedback, latency in the safety system exists (Rasmussen & Pedersen, 1984). Insufficient SA and latent error conditions have been discussed, which can contribute to an unwanted network of events or causal path, and thus, there is a need for systems knowledge of how past errors are later detected by the air engineer who suffered the error to design system defences or controls.

The nature and extent of multi-process influences on I-LED is an under explored safety field. As far as could be determined from the vast corpus of existing literature, no clear descriptions were found for I-LED associated with the unplanned and seemingly spontaneous recall of past activity. Existing literature focuses on error avoidance or proximal detection (from individual and systems perspectives) rather than the detection of errors post task completion. This could impact the organisation's safety goals (Rasmussen & Pedersen, 1984; Latorella & Prabhu, 2000). In the absence of specific safety research, a multi-process approach has been introduced that indicates I-LED is dependent on three distinct areas: the quality of PM encoding; triggering conditions in the physical environment; and SAS monitoring. This introduces a novel theoretical framework to contribute to systems thinking, which is argued to account for the total human interaction with the physical environment. Here, it is hypothesised that PM is responsible for the preparation of schema (schema selection and trigger criteria) and is linked to SAS theory for the execution, monitoring and feedback (housekeeping activity). This housekeeping function may be dependent on sensory data within the physical environment for I-LED events to occur or, at least, give rise to self-doubt that something has been missed such that the operator may be compelled to check their work (Sunderland et al., 1983; Baddeley, 1997). This appears analogous to the observed phenomenon where an aircraft engineer becomes suspicious that a completed task may be in error, from which he or she suffers the overwhelming desire to return to check their work.

Thematic analysis of DASORs facilitated the initial exploration of I-LED events by providing evidence for the observed phenomenon and applicability of the proposed multi-process theoretical framework, for which schema mismatch appears to be dependent on system cues in the physical environment for successful I-LED to occur. Yet the exact nature, extent and observable effects remain a mystery, which provides the impetus for the safety research described in this book. Here, the theoretical framework needs to be tested through naturalistic real-world studies to gain understanding of this under explored safety field from a sociotechnical stance. Dekker (2003, p. 100) suggested that understanding the mind comes from analysis of the 'world in which the mind found itself instead of trying to pry open the mind' (i.e., understanding the role of the physical environment) and is an important statement as the need for 'operational' understanding of error behaviours is recognised (Flin et al., 2008) and drives an argument for schema research to be conducted in real-world contexts. Deliberately, this positions current safety research within the realm of HFE research rather than cognitive psychology, as it is the influence of the external environment on safety behaviour that, arguably, presents the greatest safety value (Plant & Stanton, 2013b). Thus, the studies described in the following chapters will first thematically analyse narrative data from aircraft engineers who have experienced an I-LED event before moving to empirical research within the physical working environment to observe system interventions based upon multi-process findings. Ecological experiment also brings benefit with observational studies since schema are internal representations of the world for which measurement can only come from observed behaviour in the real-world contexts (Plant & Stanton, 2013b).

Exploration of the I-LED phenomenon also needs to account for the various experience levels present in naval aircraft engineers. Since trainees are naturally at an early stage of learning (Fitts & Posner, 1967), expectantly, there is likely to be greater variability in performance behaviours due to schema development. It is therefore difficult to assure stability or meaning from studying these trainees, and thus research should focus on trained aircraft engineers. Although the trained operator is not immune to error, their ability to spontaneously self-detect situational errors is a notable characteristic and needs to take account of safety behaviours and attentional mechanisms leading to I-LED.

I-LED may also enhance existing safety systems that already mitigate for the inevitability of error, thereby supporting the concept of resilience introduced earlier (Reason, 2008; Woods et al., 2010), although it is accepted that zero human error is likely to be an unrealistic safety goal (Kontogiannis, 1999; Woods & Hollnagel, 2006). Woods et al. (2010, p. 6) suggested 'people create safety in the real-world under resource and performance pressures at all levels of the sociotechnical system'. This is a view supporting both systemic and individual error contexts, which drives the need for HFE research in truly naturalistic real-world settings. This provides the opportunity to understand the nature and extent of situational error from a systems stance, for which I-LED is the effect to be observed. Without this knowledge, humans do not evolve their capabilities (Reason, 1990). Specifically, this chapter has offered a new theoretical framework upon which to explore the new safety field of I-LED and identify interventions that enhance I-LED events in aircraft engineers (and wider) using a human-focused systems perspective. However, it is accepted that achieving meaningful real-world research is challenging since strict experimental control is not possible, although the potential ecological benefit should outweigh such concerns to deliver meaningful contributions. Chapter 4 explores this challenge by exploring the role of I-LED as an additional safety control and introduces a research strategy upon which to observe I-LED events in the normal everyday workplace environment. Prior to this, Chapter 3 reviews the concern highlighted in this chapter with regards the use of the term human error in systems safety research.

3 Rationalising Systems
Thinking with the Term 'Human Error' for Progressive Safety Research

3.1 INTRODUCTION

Chapter 2 introduced the I-LED phenomenon, for which I-LED events are likely to offer further mitigation for human error effects or erroneous acts caused by system failures. Safety research presented in this book attempts to advance thinking on systems safety by researching the nature and extent of I-LED events associated with past human errors. Thus, it is necessary to authorise a theoretical position that rationalises systems thinking with the term human error, as both are central themes in the exploration of the phenomenon. Use of the term human error has been cited as outdated and should be retired in favour of a new emergent systems lexicon, since reference to human error in causation modelling can be used to infer individual blame, and therefore, wider sociotechnical causes are ignored (Dekker, 2014). It is argued in the following chapter that Human Factors and Ergonomics (HFE) research from a systems perspective requires a human-centred approach where human error effects are simply the catalyst for wider HFE analysis (Carayon, 2006) and is therefore congruent with progressive safety research. HFE analysis yields real-world application in terms of mitigating system failures through organisational control measures impacting safety behaviour at the local level; thus, the term human error should be employed, and not retired.

3.2 HUMAN ERROR

Traditional human error research often referred to situations where either the safety or the effectiveness of a task is compromised due to human failings (Reason, 1990; Hollnagel, 1993; Amalberti, 2001; Wiegmann & Shappell, 2003). There are many definitions for human error, which Chapter 2 provided a review of some of the most frequently used. Generally, though, human error describes occasions where human performance was not enacted as expected due to system-induced influences (Reason & Hobbs, 2003; Leveson, 2004; Reiman, 2011). However, the term human error has been cited as a misunderstood term that causes a focus on individual failings rather than seeking to understand the macro-ergonomic view of system failures that cause human error effects (Dekker, 2014). Applied academic research has long recognised the influence of wider sociotechnical issues that can lead to performance variability (Hutchins, 1995; Stanton & Baber, 1996; Reason, 2008; Woods et al., 2010; Cornelissen et al., 2013; Dekker, 2014; Hollnagel, 2014; Chiu & Hsieh, 2016), but it is perhaps understandable that society has favoured the focus on individual failings since they are most often the easily identifiable and explainable effect at the sharp end of safety operations rather than trying to analyse and mitigate for the system causes complicit in complex sociotechnical networks (Carayon, 2006; Flin et al., 2008; Salmon et al., 2016).

Human error is a normal by-product of performance variability, which encapsulates a range of human interactions within a given sociotechnical environment that includes non-normative and normative behaviours in the workplace (Reason, 1990; Woods et al., 2010; Cornelissen et al., 2013; Saward & Stanton, 2017). Error is the observable effect of system failures due to deficiencies in organisational safety strategies where the real causes of human error are deep-rooted in system factors such as organisational decisions, equipment design, management oversight and procedures (Woods et al., 2010; Dekker, 2014; Stanton & Harvey, 2017). To avoid the temptation to blame individuals, some safety thinking has moved to retire the term human error in favour of terms such as erroneous acts, human performance variability or system failures (Dekker, 2014). Arguably, this move is not likely to remove the temptation to conclude individual failings, and thus it risks being seen as semantics unless greater emphasis on the systems perspective is shown to benefit the organisation through reduced harm and/or increased productivity, which is covered in Chapter 9. In moving away from explaining error to understanding how system design caused a deviation from normative and expected procedures, there is a need to understand the factors that shape individual safety behaviour. HFE research requires knowledge of human performance-shaping factors or error-promoting conditions so that the wider societal factors and the technical environment can be matched and controlled reliably to mitigate for human performance variability (Reason, 1990; Leveson, 2004; Hollnagel, 2014; Rasmussen, 1997; Carayon et al., 2015). This is an area that characterises operator competence, which is an essential element of a resilient system, and is discussed further in Chapter 8.

HFE is concerned with human interactions within systems; thus, Fedota and Parasuraman (2009) also rightly challenged this new paradigm shift away from human error as overly focussing on the organisational factors, which risks ignoring important factors associated with human behaviour within a specific context. This is a concern held by Plant and Stanton (2016), for which they highlighted the capability of the schema-based perceptual cycle in linking human elements of cognitive behaviour with the operating environment. When human error is discovered in the workplace, it is a trigger for HFE analysis of the system deficiencies that generate error-promoting conditions causing failures in the system, which can lead to undesirable risks such as an accident or reduced performance. This is the starting point for HFE analysis, not the trigger to blame individuals (Stanton & Baber, 1996; Reason & Hobbs, 2003; Reiman, 2011; Saward & Stanton, 2015a). The term latent error condition refers to safety failures that become embedded in the system, which can impact future safety performance, that is system-induced conditions that promote errors that pass undetected and then lie hidden in the sociotechnical system (STS; Reason, 1990). The detection of past errors is an essential element of achieving system safety (Rasmussen & Pedersen, 1984; Reason, 1997; Shorrock & Kirwan, 2002; Wiegmann & Shappell, 2003; Flin et al., 2008; Aini & Fakhru'l-Razi, 2013). Thus, it is argued that human factors experts analysing error events should not avoid using the term human error, provided it is used to signpost residual error effects caused by system influences that require attention; it is an effective term to trigger HFE analysis of the wider causes in complex STSs, both in design of safe systems and implementation of control measures for the restoration of safety when a system failure occurs. Indeed, few go to work to deliberately cause an accident; thus, any attribution of individual blame is a failure to understand systemic causal factors where accountability needs to be balanced against learning from system failures (Reason, 1990; Dekker, 2014).

To be assured of where the term human error sits in progressive safety research, the role of systems thinking and the function of HFE in preventing the escalation of causal paths to an accident need further consideration.

3.3 SYSTEMS THINKING

Systems thinking transfers the emphasis from individual human failings to understanding the living network of all sociotechnical factors that can cause system failures (Leveson, 2011). The systems view favours this macro-ergonomic approach to safety rather than the micro-ergonomic lens to avoid focussing on individual human failings (Zink et al., 2002; Murphy et al., 2014). Thus, macro-ergonomic analysis explores the sociotechnical interaction between elements comprising humans, society, the environment and technical aspects of the system, including machines, technology and processes (Reason & Hobbs, 2003; Woo & Vincente, 2003; Walker et al., 2008; Amalberti, 2013; Wilson, 2014; Niskanen et al., 2016). These networks can be complex in terms of the number interactions between systemic factors such as tools, equipment, procedures, decision making, operator training and experience, and operating contexts (Edwards, 1972; Reason, 1990) and is where progressive safety strategies recognise that every element of the STS contributes to the organisation's safety goals through specific roles, responsibilities, relationships and safety behaviours (Leveson, 2011; Dekker, 2014; Plant & Stanton, 2016). A safety failure occurs when there is inadequate control of the sociotechnical factors (across networks), which impacts human performance (Woods et al., 2010). Each has dependencies on the other in the network, so when things go wrong, it should be recognised as a failure of the system as a whole that has resulted in a hazardous condition due to design-induced factors and not individual errors (Stanton et al., 2009a). The aircraft engineer is one element in a complex sociotechnical system of aircraft maintenance where system deficiencies can cause failures, borne out in observable error effects. Chapter 2 described the complex sociotechnical environment of naval aircraft maintenance, but complexity is not a stranger to other contexts such as commercial aviation, energy suppliers, transportation, logistics and health care where complexity comes not only from the human–machine networks but wider geographical, temporal, cultural, socio-political, regulatory, technological and economic dimensions that place extreme demands on human performance in these ever-changing workplace networks

(Blavier et al., 2005; Carayon, 2006; Leveson, 2011; Hollnagel, 2014). Arguably, over reliance on the macro-ergonomic approach can mask the detection of all human factors impacting safety success within these networks by the very nature of the complexity involved. No universal method exists for the macro analysis of sociotechnical systems (Kleiner et al., 2015), which generates scope for significant variance in safety-related control strategies.

Organisational or system accidents occur when there is insufficient awareness of the risks of safety failures and/or there is ineffective control of the interacting component hazards within the STS (Leveson, 2004). Critically, it has been highlighted that an organisational accident is rarely the result of one error event, since it is more usually the networking of more than one error event or system failure that creates a causal path to the system accident (Perrow, 1999; Amalberti, 2013). But despite a systems approach to the design and implementation of a safe system of work, the accident can still come as a surprise to safe or ultra-safe organisations (Hollnagel, 2014). Arguably, the macro-ergonomic approach risks missing important human factors for the delivery safe systems, which is a void that is often bridged through the heroic abilities of operators in the system. Here, Reason (2008) views humans as 'heroes' where safe behaviour exists that adapts to system failures to enact a successful recovery, which supports safety resilience. Similarly, Hollnagel's (2014) modelling of accident causation highlighted Safety II events where the adaptive capability of human operators can locally overcome or avoid system failures in the workplace (heroic recoveries), whilst his Safety I analogy refers to error avoidance and capture through the planning and delivery of effective safety controls aimed at defending against identified hazards.

Shappell and Wiegmann (2009) found recommendations from National Transportation Safety Board (NTSB) aviation accident reports often focussed on organisational changes or design improvements rather than taking a more balanced sociotechnical approach that includes recommendations to improve the safety behaviour of operators during normal operations. Their point being that accident rates in aviation have largely stabilised over recent years, thus if the adage is true that there are no new accidents, just new operators ready to suffer a system failure, then more needs to be done to design system interventions that account for the inevitability of performance variability by capitalising upon heroic recoveries. Changing or enhancing safety behaviour needs to form part of an enduring system solution, as the control measures to affect the change need to be integrated within the overall safety design. Arguably, I-LED events offer the potential to enhance safety behaviour by maximising the use of system cues to mitigate for inevitability of human performance variability, and therefore, I-LED is a system solution to counter the risk of a causal path to an organisational accident.

Not all examples of normal operations can be bounded by a procedure, as complex operating environments can be highly dynamic, which generates novel and unexpected safety issues. I-LED events can perhaps mitigate for error effects caused by unexpected safety challenges. Indeed, safety is itself the effect from successful system interactions (Leveson, 2004), which is likely to be routed in effective management of system hazards from planned safety through to the reality of everyday normal operations. If humans can exhibit heroic recoveries, and this truly mitigates for exceptional circumstances, then progressive safety research should re-balance the systems view by ensuring more is understood about individual (and team) safety behaviours. I-LED is believed to contribute to this more complete view of safety if integrated with existing system controls, and thus, research in this area is believed to be progressive and underwrites to the macro-ergonomic view. The systems view is progressive thinking, but it must also pay due consideration to how individual failings can be avoided through local responses to everyday safety challenges not accommodated in the safety design. This is analogous to the holes in Reason's (1997) Swiss cheese that have not all been 'plugged' by a safety control such as training or a procedure, yet operators can often locally detect persistent and erroneous holes affecting safety performance and respond effectively (Amalberti, 2013). Expectantly, the system must define and achieve a level of operator competence to support successful I-LED events, which is discussed in Chapter 8.

Effective safety strategies therefore need to plug holes to offer sufficient control over hazards that pose a risk to system safety and offer effective 'as done' safety at the sharp end of normal operations (Flin et al., 2008; Morel et al., 2008; Amalberti, 2013; Hollnagel, 2014). The studies in the following chapters show that I-LED interventions support 'as done' safety activity in the workplace and are practicable and sufficiently flexible to accommodate exceptional occasions caused by system failures (Reason, 1990). This contributes to workplace safety strategies provided I-LED interventions are integrated within the organisation's safety system to ensure training, time to conduct the intervention, and the availability of context dependent cues are available in the workplace. This is argued to offer a progressive application of real-world safety, which is discussed further in Chapter 8. Notably, I-LED research can perhaps help counter other potential consequences of latent error conditions, which are not safety related, such as overall system performance, social-economic gains, and political and reputational value (Kleiner et al., 2015) that may form part of wider resilience against organisational decision that create potential opportunities for system failures.

3.4 SUMMARY

Progressive safety research requires a systems view but equally must not forget that the individual human is at the fore of the safety solution (Flin et al., 2008), which drives the need to assure operators are competent to meet the safety expectations of the safety system ('as designed'; Hollnagel, 2014). A competent operator enhances safety resilience through their heroic recoveries by detecting and recovering from human error caused by system failures. The importance of human performance factors and safety behaviour in the workplace drives the argument that the system needs to ensure the competence of its operators can achieve the safety expectation of the safety system. Effectively, human capabilities must be matched to the operating environment such as a trained and experienced aircraft engineer or pilot. Both are specially selected for certain performance attributes that are needed to bound performance variability within the environment of aircraft maintenance or the flight deck. A pilot may not make a successful engineer and vice versa. Thus, matching human performance is essential. Understanding safety behaviours and how they can be managed at the sharp end to counter human error effects by plugging holes in the safety system is one of the main elements of the STS that should contribute to an organisation's safety goals (Flin et al., 2008; Leveson, 2011; Hollnagel, 2014).

The term human error has been argued to remain meaningful when conducting HFE research from a systems perspective and therefore does not need to be excluded from progressive safety research such as I-LED. As with any lexicon, terms must be used correctly, for which human error is simply a subset of macro-ergonomic systemic factors that flags the requirement for HFE analysis of systemic factors. As such, the term is used universally in the current exploration of the I-LED phenomenon where the ability for individual operators to create safety through their own I-LED behaviour is as much a system safety solution as the wider HFE design and implementation of systemic defences or controls such as procedures, training, equipment, the operating environment and management of safety at the organisational level. I-LED is therefore argued to be congruent with the systems thinking where HFE analysis of human error is argued to remain relevant and meaningful in progressive safety research provided the term is used from a systems perspective to describe performance variability effects caused by system failures and not overly focus on individual failures. This position is returned to throughout the following chapters when considering how to make the safety system safer. The notion of competence and resilience is discussed in Chapter 8 when addressing the role of I-LED interventions in optimising resilience, whilst the following chapter explains the research strategy to observe I-LED events based upon the theoretical perspectives discussed in this chapter.

4 Observing I-LED Events in the Workplace

4.1 INTRODUCTION

The previous chapter described the potential risk of harm to people and/or equipment posed by a system failure when maintenance error is not detected proximal to the task. An undetected system failure leads to a latent error condition that can later give rise to an unwanted outcome, such as degraded system performance or, at worst, an accident. Rarely is an organisational accident the result of a single cause (Perrow, 1999; Amalberti, 2013); thus, in terms of undetected errors, a latent error condition can network with other safety failures to create a causal path or chain of events within the Sociotechnical System (STS) that can cause harm. An effective safety strategy to counter or defend against hazards becoming a system failure or transitioning to a latent condition is to design Human Factors and Ergonomics (HFE) safety controls to target system hazards so that they are properly managed without risking a safety disturbance or safety failure (Carayon, 2006; Leveson, 2011; Johnson & Avers, 2012). Application of safety controls throughout the STS to avoid harm, as opposed to aspiring to zero system failures, is characteristic of a resilient system. This requires a systems approach that comprises HFE-designed safety controls targeting the network of hazards, system failures and potential error-promoting conditions that occur within complex operating environments (Reason, 2008; Woods et al., 2010; Leveson, 2011; Amalberti, 2013; Hollnagel, 2014). The I-LED phenomenon introduced in Chapter 2 is thought to act as an additional system control that aids an individual's detection of their past error by exploiting system cues to trigger the recollection of past activity. Thus, I-LED is arguably representative of an additional safety control that helps defend against networks of safety failures and error-promoting conditions that form a causal path to an accident, thereby contributing to resilience within the safety system. To explore the I-LED phenomenon, this chapter describes the research strategy needed to explore the phenomenon, which includes the methodology and series of linked studies designed to observe I-LED events in real-world workplace environments.

4.2 RESEARCH STRATEGY

4.2.1 Target Population

The target population exists within the operating context of UK naval aircraft maintenance where around 1700 naval air engineers are employed, which includes junior operatives through to senior supervisors, being male and female of ages 18–50 years. As the Royal Navy (RN) is an equal opportunities employer, military personnel are all British Commonwealth citizens of broad ethnicity. Participants in each study have been sampled from this population, taking care to ensure samples do not overlap such that data from adjacent studies risked contamination. Sample groups were selected and sized to be meaningful and statistically valid as required by each study. Sampling by operator group (operative or supervisor) accounts for a broad range of experience levels. This population was chosen since the lead author is a serving Royal Navy Air Engineer Officer (AEO) and therefore provided a deep knowledge of the operating context, free access to the target population and safety databases where needed, resulting in a data-rich environment to explore. Whilst knowledge and expertise in a subject area are needed to exploit the maximum information from the available data, the authors recognised the need to ensure this deep knowledge did not inadvertently bias the research, and thus, various strategies are used in each study to counter any effects. This included testing for inter-rater agreement, review of electronic recordings where used, and studies designed based upon theory and evidence from other studies to derive question sets and analyse data (Cassell & Symon, 2004; Schluter et al., 2008).

4.2.2 IMPORTANCE AND CHALLENGE OF NATURALISTIC REAL-WORLD OBSERVATIONS

To explore the nature and extent of I-LED in the workplace and its contribution to safety, everyday routine safety behaviour needs to be observed since naturalistic research allows successful error detection and recovery to be observed in response to real-world scenarios (Saward & Jarvis, 2007). Unlike highly procedural environments found on a flight deck or in process control, it is arguably more difficult to observe everyday error behaviours of the aircraft engineer due, in part, to the complex, dynamic and multiple operating environments experienced by the aircraft engineer when carrying out day-to-day maintenance tasks (Prabhu & Drury, 1992; Latorella & Prabhu, 2000; Rashid et al., 2010). For example, the cockpit environment benefits from flight data and cockpit voice recordings, which provide a rich source of real-world data (Flin et al., 2008). This facilitates an accurate account of error detections. Conversely, nothing comparable exists within the aircraft maintenance environment, so errors are not always apparent when observing the naturalistic environment and can therefore pass undetected, leaving it very challenging to observe I-LED events. Notably, Flin et al. (2008) argued that understanding of error behaviour could only come from the assessment of operational behaviours in context, whilst Woods et al. (2010, p. 236) believed *research on how individuals and groups cope with complexity and conflict in real-world settings produces the necessary insight on how to approach error.* Dekker (2003, p. 100) suggested that understanding the mind comes from analysis of the *world in which the mind found itself instead of trying to pry open the mind* (i.e., understanding the role of the physical environment) and is an important statement, as the need for 'operational' understanding of error behaviours is widely recognised and drives an argument for schema research to be conducted in real-world contexts. Deliberately, this positions current research within the realm of

HFE research rather than cognitive psychology, as it is the influence of the external environment on error behaviour that, arguably, presents the greatest safety value (Plant & Stanton, 2013b). Thus, future studies will first thematically analyse narrative data from aircraft engineers who have experienced latent error detection before moving to empirical research within the physical working environment to observe I-LED interventions designed using multi-process findings. Ecological experiment also brings benefit with observational studies since schema are internal representations of the world for which measurement can only come from observed behaviour in the real-world contexts (Plant & Stanton, 2013b).

Much enquiry has been made of error avoidance and proximal error detection (and recovery) where mostly empirical analysis involving carefully controlled laboratory experiments has been conducted (Allwood, 1984; Kanse, 2004; Thomas, 2004; Flin et al., 2008; Kontogiannis, 2011; Patel et al., 2011; Wilkinson et al., 2011). Understanding safe behaviour through local safety practices can only come from assessing operational behaviours in context. However, it is recognised that strict experimental control is not possible in naturalistic studies, although the advantage of naturalistic studies is that it avoids the bias of artificial controlled experiments that can erode the ecology of findings. Studies have also shown the nature of the operating context is not easily replicated in simulated experiments due to the complexity of sociotechnical interactions (Prabhu & Drury, 1992; Latorella & Prabhu, 2000). Sellen et al. (1996) found conducting their experiments in a work environment provided a level of ecology that yielded important sociotechnical factors such as the influence of competing activities/tasks, time and physical locations, although they highlighted the challenge of tracking performance in dynamic real-world environments. Thus, it might not possible to faithfully emulate realistic system contexts or cues in a laboratory setting, but the potential ecological benefits are widely recognised to outweigh any potential concerns to deliver a meaningful contribution to safety knowledge (Norman & Shallice, 1986; Dekker, 2003; Kvavilashvili & Mandler, 2004; Flin et al., 2008; Reason, 2008; Finomore et al., 2009; Woods et al., 2010). As this may enhance resilience to the inevitability of human error, the authors were confident that a real-world study in the natural workplace was needed to investigate the phenomenon, but two areas needed to be addressed: use of appropriate theory against which to assess I-LED performance; and selection of research methodologies that facilitate effective data gathering from the workplace.

4.2.3 THEORY UPON WHICH TO OBSERVE I-LED EVENTS

The extensive review of literature in Chapter 2 found little research on I-LED events occurring post-task completion. In the absence of current theories, the case was argued for a multi-process approach to systems research that combines theories on prospective memory (PM), supervisory attentional system (SAS) and schema theory to observe I-LED events. Combining several theories to achieve a multi-process research strategy suggests it is unlikely that a single factor (independent variable) would be observed, and therefore, multiple sociotechnical factors with interrelated determinants would need to be considered in each study.

When reviewing theory upon which to explore I-LED events, it was noted that the schema theory element of the theoretical multi-process framework has been challenged as to whether it is appropriate theory for HFE research. Bartlett (1932) introduced schema as information represented in memory about our external world comprising objects, events, motor actions and situations. Norman (1981) defined this information as 'organised memory units' for knowledge and skills that are constructed from our past experiences, which we use to respond to information from the world. This human interaction with the world is characterised by the Perceptual Cycle Model (PCM), which represents human actions based on the cyclic interaction of schemata with perceptions of the world. The model can be largely automatic for skilled operators using well-learnt skills (Norman & Shallice, 1986), which includes the ability to spontaneously detect past errors, as discussed in Chapter 2. Here, it is the lack of definition and understanding of the internal workings of the mind plus the absence of a unitary value to measure schema performance that has been criticised (Mandler, 1985; Smith & Hancock, 1995; Endsley & Robertson, 2000). Plant and Stanton (2016) also recognised the inherent difficulties in measuring schema performance in the external environment but argued the use of schema theory in HFE research is more concerned with how we interact with the world rather than specific knowledge of genotype and phenotype schema, which are themes described in Chapter 5. Dekker (2003) supports this position with his argument that safety gains come from understanding the world around the mind as opposed to its internal workings. It is argued that knowing internal memory units exists and that they possess a hierarchy for correct sequencing is sufficient, rather than trying to unlock detailed knowledge of exactly how the mind works. Applying schema theory in the knowledge that an operator matches tasks (PM element) with actions by monitoring (SAS) cues in the world (PCM) is believed sufficient for I-LED research. Plant and Stanton (2016) continue to highlight much research in Situational Awareness (SA) is based upon schema theory, which has resulted in significant gains in safety research. Stanton & Walker (2011) argued schema theory has been tested over time and shown to provide sufficient account of how operators interact with the sociotechnical environment, and it can also help identity where system failures can occur. Thus, schema theory, as part of multi-process approach to I-LED research, is argued to be appropriate theory.

4.2.4   Linked Studies Approach to Research

The second challenge to real-world research is selecting research methodologies that facilitate effective gathering and analysis of data from the natural workplace. To help ensure quality data were captured, and to remain flexible to emergent findings, a staged approach to research using a mixture of methodologies is employed in a series of linked studies (Trafford & Leshem, 2008). Strategically, this flexible approach facilitates the application of various research paradigms and data collection instruments that can be applied when observing the target population over a protracted period. This strategy also safeguarded against any study that failed. Emergent findings from each study were used to guide research in terms of direction for subsequent studies. This required continued engagement with current literature for the iterative development of hypothesis and theories; expectantly, this has matured thinking on the phenomenon as each study reported its findings.

To provide direction for each I-LED study, research was framed around the aim and Objectives 2, 3 and 4 stated in Chapter 1 to construct three linked observational studies to investigate the I-LED phenomenon:

- *Objective 2: Study 1 (Exploratory group sessions).* To 'baseline' research and facilitate analysis of emergent themes relating to the proposed phenomenon exploratory group, sessions were conducted, which included the administration of a questionnaire designed according to multi-process theories. Group sessions allowed an explanation of the research to be given and to address any questions with the questionnaire in a non-directive manner to ensure completeness of individual responses (Oppenheim, 1992). This approach was also selected to collect a large amount of data efficiently whilst fully inducting the authors in the concept. Data were analysed thematically (Robson, 2011) to highlight significant findings and to further conceptualise and contextualise the proposed phenomenon before identifying themes/hypothesis to test in subsequent studies.

- *Objective 3: Study 2 (Workplace self-report diaries).* A self-report diary was used to collect data on everyday I-LED events occurring in the workplace. Whilst capturing individual perceptions, the advantage of self-report diaries as a research instrument is that it can be used ecologically to observe the subjective phenomenon by reporting on specific events or tasks present in everyday activity without intrusion but with adjacency and detail (Reason, 1990; Cassell & Symon, 2004; Robson, 2011). The diary recorded I-LED observations over two months and was constructed using Critical Incident Technique (CIT), which is the general term that describes the process of capturing important incidents recorded by a participant (Flanagan, 1954); an incident being *any observable human activity that is sufficiently complete in itself to permit inferences and predictions to be made about the person performing the act.* The term critical simply refers to a significant I-LED event reported by the participant. To help determine whether the target population for the diary study exhibited normal cognitive behaviours, literature was reviewed for an appropriate instrument to employ. The Cognitive Failures Questionnaire (CFQ) scores an individual's propensity for everyday cognitive failures using 25 questions scored 0–100 against a five-item Likert coding, where a high mean score (>51) indicates a propensity for cognitive failures (Broadbent et al., 1982). The CFQ was used to confirm normal cognitive behaviours in the previously unstudied cohort of naval air engineers but not to judge individual cognitive performance whilst also providing a benchmark upon which to authorise research findings against wider populations. Data from the self-report diaries were analysed thematically to highlight significant findings, thereby providing direction for Study 3.

- *Objective 4: Study 3 (I-LED interventions).* Study 2 indicated the application of targeted I-LED intervention techniques that draw upon system cues are likely to enhance the effective detection of past errors. The study also argued that targeted interventions are especially important for simple everyday habitual tasks carried out alone where perhaps individual performance variability is most likely to pass unchecked if there are organisational deficiencies in system defences. Thus, appropriate I-LED interventions (for a specific safety context) are likely to offer further resilience against human performance variability. Study 2 reported any intervention needed to be deployed within a real-world context to operationalise and assess its benefit. Thus, Study 3 has been designed to advance knowledge on I-LED by observing the utility of I-LED interventions within the workplace determined from Study 2. To help provide direction on the practicable interventions to be observed in the study, additional qualitative data collected during Study 1 was accessed. This analysis guided the selection of appropriate I-LED intervention techniques to be used in the study. The findings from Study 3 are used to answer Question 3 and therefore complete the component parts needed to address the main research question.

4.3 SUMMARY

I-LED events are believed to act as an additional control measure (Safety II) that can be integrated within an existing safety system that is applied locally by operators. To conduct real-world observation in the natural working environment is challenging when compared to observations made in a laboratory setting, where strict experimental control can be achieved. However, advances in safety research come from observing real-world behaviours during normal operations, and this advantage has been argued to outweigh the challenges of naturalistic research. Part of this challenge is the application of appropriate theory and method selection to yield meaningful results from the population of naval aircraft engineers. Thus, the design of the current safety research allows sufficient flexibility via series of linked studies using mixed methods. This strategy also accommodates changes to the design if emergent findings materialise that require a different methodology or revision to the observed phenomenon based on real-world findings. The next chapter introduces the first study, which uses group-administered questionnaires to determine the presence and nature of the proposed I-LED phenomenon in the target population.

5 A Time and a Place for the Recall of Past Errors

5.1 INTRODUCTION

To observe the I-LED phenomenon using a systems approach, Chapter 2 proposed a multi-process theoretical framework, combining theories on Prospective Memory (PM), Supervisory Attentional System (SAS) and schema theory. The multi-process framework argues the PM element forms intent for a task to be carried out in the future (Baddeley & Wilkins, 1984; Kvavilashvili & Ellis, 1996; Blavier et al., 2005; Dismukes, 2012), which creates a 'to-do' list or markers in the mind (Sellen et al., 1996; Marsh et al., 1998; Van den Berg et al., 2004). The schema theory element describes information represented in memory about our knowledge of the world with which we interact (Bartlett, 1932). Schemata are developed over time through past experiences and account for the knowledge and the skills that we apply to everyday situations in response to external sensory cues (Norman, 1981). This is an iterative process where repeated exposure to similar situations leads to highly developed schemata that are associated with an experienced operator who becomes well-practised at particular tasks (Bartlett, 1932). Developed schemata that form internal memory structures, which can be accessed to respond to a particular task, are known as genotype schemata, whilst phenotype schemata refer to the actual response when executing a task (Neisser, 1976; Reason, 1990). This is an important distinction in schema theory as it is the phenotype schema that manifests as the observable effect when assessing human activity, and therefore, it this phenotype schema behaviour that is of particular relevance when exploring I-LED. Chapter 4 highlighted some of the criticisms of schema theory due to its under specification as a theorem that lacks unitary value to conduct measurements. However, a review of literature in Chapter 4 found schema theory to be a useful theoretical approach to observing human interactions with the world, which can be used to help understand sociotechnical factors influencing safety behaviour.

Schema theory is further characterised by the Perceptual Cycle Model (PCM). The model describes a cyclic relationship between schema selection and sensory cues in the external world that trigger human actions (Neisser, 1976). Thus, the PCM is argued to sit within the multi-process framework to account for human interactions with the world. The PCM and schema theory indicate that only the highest-level schema needs conscious control for subordinate schema to trigger autonomously (Norman, 1981; Mandler, 1985; Cohen et al., 1986; Reason, 1990; Baddeley, 1997). For example, an engineer may be tasked to flight service an aircraft. This requires the conscious selection and triggering of the high-level schema for aircraft servicing, which in turn necessitates the triggering of subordinate schemata that are automatically associated with the high-level task such as opening a panel, using tools, checking levels, replenishing fluids and so on. It is argued the engineer is more likely to detect the high-level schema being in error due to forgetting to carry out the entire flight servicing task, as opposed to one of the many largely autonomous and discrete subordinate schemata. A latent error condition is generated when the internally selected genotype schema is either not triggered or triggered, but the phenotype schema, which manifests as the actual human action, does not match the action required for the task. In effect, the failed trigger or genotype/ phenotype mismatch has not been detected by the PCM proximal to the error event (for comprehensive coverage of PCM theory refer to: Bartlett, 1932; Neisser, 1976; Norman, 1981; Mandler, 1985; Stanton et al., 2009a; Plant & Stanton, 2013a).

The SAS element of the multi-process framework, proposed by Norman and Shallice (1986), is attributed the attentional mechanism that continuously monitors the external world for sensory cues that trigger schemata and which provides task feedback within the PCM schema-action-world cycle (Smith, 2003; Einstein and McDaniel, 2005). Thus, it is this SAS element that regulates the cancelling of completed tasks on the 'to-do' list. This task feedback also updates established genotype schema and enables the development of new schema if the task was novel (Reason, 1990; Plant & Stanton, 2012). Multi-process theory constructed in Chapter 2 coined this activity as the schema housekeeping function within the PCM where intended actions are monitored for completion, and feedback from the action also facilitates learning and the acquiring of experience.

Chapter 3 discussed the need to approach human errors effects as the catalyst for understanding system failures that caused the error. The system perspective, therefore, considers the world in which people are immersed in and which drives safety behaviours from a human-centred approach. The PCM accounts for human error from the systems perspective, as shown in Figure 5.1, which has been adapted from Neisser (1976) and Smith and Hancock (1995) to highlight the elements of multi-process theory introduced in Chapter 2. Here, the sociotechnical world provides the system cues that trigger actions, from which the PM element was argued to be responsible for the creation of a 'to-do list' of tasks, described in Chapter 2. The PM selects required schemata (cognitive information) to carry out the action as part of this activity. Task execution occurs when the PM directs behaviour based on samples of the world by the SAS (Plant & Stanton, 2012). Execution of this activity requires the bottom-up (BU) processing of external data from sensory inputs with top-down (TD) prior knowledge (schemata) to project a response (Neisser, 1976; Cohen et al., 1986; Plant & Stanton, 2013a). This processing of the external environment with internal pre-existing schema to achieve the required response captures elements needed for Situational Awareness (SA) and is essential for the execution of tasks without error (Plant & Stanton, 2013a). When errors occur, the continual monitoring by the SAS to assess task performance and schema reinforcement/learning through schema housekeeping (Saward & Stanton, 2015a) also provides the opportunity for I-LED events.

The combination of multi-process theory and the PCM shown on Figure 5.1 is argued to encapsulate cue dependency for the triggering of schemata, which is dependent on the effective cognition of sensory cues distributed throughout the

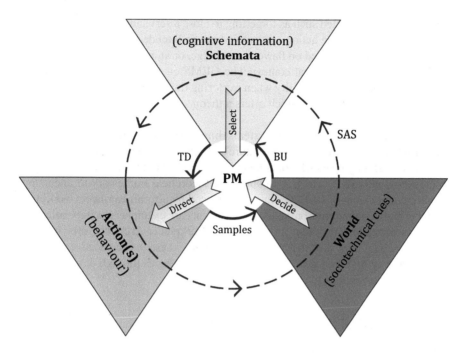

FIGURE 5.1 Multi-process PCM. (Derived from Neisser, U. 1976. *Cognition and Reality: Principles and Implications of Cognitive Psychology.* San Francisco, CA: Freeman; Smith, K. & Hancock, P.A. 1995. *Human Factors*, **37**, 137–148.)

external environment (Hutchins, 1995). Thus, the relationship between the mind and the external world is important, for which distributed cognition theory argues cognitive cues are distributed throughout all sociotechnical environments in which the mind finds itself (Hutchins, 2001). This helps to understand that human actions are dependent on cues distributed throughout the external environment and leads to the concept of Distributed Situational Awareness (DSA), which is needed for effective human performance (Stanton et al., 2015). This dependency on external cues to trigger schema action has been investigated in PM studies of human error proximal to task activity, which have reported that error avoidance relies on the successful cognition of system cues (Grundgeiger et al., 2014). Thus, it is hypothesised that I-LED is dependent on sensory cues distributed throughout the external environment. Further, since PCM theory describes a schema-action-world cycle (Plant & Stanton, 2013a), time must be associated with the frequency of the cycle; thus, it is hypothesised time is also a factor in I-LED.

To observe the phenomenon of I-LED, a robust method is needed to categorise the actions of the aircraft engineer, for which Reason's (1990) Generic Error Modelling System (GEMS) offers an established and widely accepted method. His model shown in Figure 2.3 describes skill-based action errors in terms of slips and lapses with rule/knowledge-based action errors described as mistakes. A slip is a well-practised task that is carried out in error unintentionally such as dropping a screwdriver (motor skill), whilst a lapse also involves a well-practised or familiar task but where the required action is omitted such as forgetting to close a servicing panel on an aircraft. Mistakes occur when the action was carried out as intended, but the wrong plan of action was formulated based on flawed knowledge, or an incorrect rule was selected for the specific sociotechnical context. The GEMS categorisation system has been argued in Chapter 2 to have utility when exploring the relationship between external triggers and error behaviour, which offers a theory-driven approach when exploring I-LED events.

The aim of this study is to apply systems thinking to the multi-process framework introduced in Chapter 2 to explore the nature and extent of I-LED amongst a population of naval aircraft engineers who operate in the aircraft maintenance context. The aim progresses Objective 2, which is to apply the theoretical framework to understand the nature and extent of I-LED events experienced by aircraft engineers working in everyday maintenance environments. I-LED events are examples of safety behaviour that helps mitigate for system failures in the workplace. Arguably, this approach supports the paradigm shift in addressing error from a systems perspective and offers the opportunity to enhance organisational safety strategies through development of practicable interventions to enhance I-LED.

5.2 METHODOLOGY

5.2.1 Questionnaire Design

The questions in Table 5.1 were constructed according to the multi-process approach to I-LED described in Chapter 2 and are therefore based upon theories for PM,

---

TABLE 5.1
Group Session Questions

Category	Question
Maintenance task	1. What was the task to be carried out?
	2. Was this your only task to be carried out? (Yes/No)
	3. How was the task identified (i.e., were you asked, was it part of a process, self-identified, etc.)?
	4. Where were you when tasked (i.e., in AMCO, hangar, working on an aircraft, etc.)?
	5. Did you carry out the task immediately or was it carried out at some point later?
	6. What was the time of day when tasked?
Error event	7. Please describe your error.
	8. What was the time lapse between when you were tasked and the error?
Latent error detection	9. Please describe how you became aware of your error.
	10. Where were you at the time you became aware of your error?
	11. What was the time lapse between the error and your detection?
	12. Have you ever been concerned that a past task was in error but found it not in error when checking (please give example)?
	13. How do you think the self-detection of past errors could be improved?

attentional monitoring and schemata. Quantitative and qualitative data were collected for the hypothesis by structuring the questionnaire around three main areas: the maintenance task; error event; and detection of error post-task completion. The measurement ability of each question was assessed against multi-process theories through participant feedback and analysis of questionnaire responses during piloting. Since questions reflected the proposed phenomenon and were structured to explore the hypothesis, this also offered a degree of construct validity and a theory-driven questionnaire that was believed to be sufficiently comprehensive to capture evidence for the exploratory nature of the current study (Oppenheim, 1992).

To protect questionnaire responses from contamination due to participants conferring or from individuals biasing other participant's views, the authors carefully controlled group dynamics. For example, the authors were ready to interject if any strong views were raised in the group or specific details of a personal error event were put forward during the group session. Conversely, to help ensure researcher immersion did not influence participants, the group sessions were recorded digitally and reviewed to check the integrity of the process and that the authors had not I-LED the participants (Cassell & Symon, 2004). This was achieved by reviewing the recordings with an Air Engineer Officer (AEO) who was not involved with the study. An engineer was used so that technical terms and the context would be understood. All recordings were found to be consistent and did not contain any strong views or leading comments that could potentially bias a group. Thus, these recordings were used for quality control only and were not subject to further analysis.

5.2.2 Sampling Strategy

To achieve a representative sample within the resources available for the current research, RN units consisting of naval aircraft engineers at helicopter squadrons and those on various maintenance courses were identified to give a sample stratified across the target population according to geographic location, type of work and experience. Within the resources available for the study, a total of 68 aircraft engineers were identified for the study from a target population of approximately 1700 (4%): eight engineers for piloting; 48 for data collection; and a further 12 to help assure thematic saturation and ameliorate participant error (Robson, 2011). This strategy required 17 group sessions to be conducted with four participants in each group (including two groups for piloting); organised by groups of four operatives and four supervisors to give a broad range of aircraft experience and ages. Groups of four were chosen to generate sufficient discussion but not so large that some individuals did not have a chance to offer a contribution or ask a question. The size of the groups also minimised the impact of the research on the aircraft maintenance unit, whilst the overall samples size provided a manageable amount of data for analysis. To identify individual participants within each unit, local management were asked to conduct a simple random sample to identify eight air engineers from their available manpower. As the sample from each squadron was very small, local management were asked to conduct a simple 'raffle' approach by selecting names 'out of a hat' (Rowntree, 1981). An additional two engineers were also requested to allow for any individuals who chose not to take part in the research or were not available on the actual day of the group session.

5.2.3 PILOTING

The administrative process and a draft questionnaire were piloted using two group sessions (operatives, n = 4 and supervisors, n = 4) that were representative of the population. Each group was asked to provide feedback based on Wiegmann and Shappell's (2001) guide for an effective taxonomy, commenting on: the comprehensiveness of the written participant information and questionnaire; diagnostic capability of the questionnaire to determine if questions were sufficiently wide-ranging and appropriate to capture everything they wanted to say about their example of a latent error condition; and how usable they perceived the questionnaire to be. This resulted in several revisions to the questionnaire and changes to the administrative process such as removing ambiguity in some questions and simplifying sentences. General readability of a Participant Information Sheet (PIS) was also assessed against the Flesch reading ease score. The PIS shown at Appendix B was given to participants, which included written information to explain the research and benefit of their contribution, scored 42.7 and thus was slightly difficult to read. Feedback from piloting helped achieve a final score of 60.1, indicating plain English for adult reading.

5.2.4 DATA COLLECTION PROCEDURE

Group sessions were conducted to introduce the concept of I-LED to participants and administer the questionnaire. This allowed the authors to explain the overall aim of the research and clarify any queries with the questionnaire in a non-directive manner to ensure completeness of individual responses (Oppenheim, 1992). Local engineering management were approached in advance for consent to conduct the safety research, and selected individuals were notified one week prior to each group session to explain the purpose of the research and how their data would be used. This was achieved via a participant information sheet (Appendix B) that was provided in advance of the group session. After reading the information, individuals were free to opt out, and another engineer from the manpower list was selected. Participants were asked to come prepared with an example of an everyday latent error condition that happened to them at work that, to reduce the potential impact from memory decay, needed to be a recent example (within the last 12 months). Each session was conducted within the normal workplace and consisted of an initial brief followed by a short period of group discussion to help understand the concept of I-LED and to provide an opportunity to ask questions before completing the questionnaire. Overall, the data collection process took three weeks, after which collected data were cleaned for errors and anonymity before analysis (Oppenheim, 1992).

5.2.5 DATA ANALYSIS

Descriptive statistics were produced for the responses to the questions shown in Table 5.1. To test for the strength of association between category variables, several contingency tables were constructed. Pearson's chi-square test was applied to each contingency table unless the frequencies were lower than five, in which case Fisher's exact test was used. Timing data were correlated using Spearman's Rho non-parametric test as data were ranked according to reported mean times. The analysis of participant narratives from Questions 9 and 13 shown in Table 5.1 was conducted using thematic analysis (Robson, 2011). Themes were generated from the participant responses shown in Appendices C and D, then compared 'horizontally' with all participants to identify any global themes (Cassell & Symon, 2004; Polit & Beck, 2004). An iterative constant comparison approach was used to review narrative data to exhaust emergent themes until thematic (theoretical) saturation was achieved (Glaser & Strauss, 1967). This required comparison with a further 12 questionnaires completed through additional group sessions to help assure thematic saturation. To test for inter-rater agreement, an independent assessor was used who conducted a 100% review of the data. Cohen's kappa was calculated on the frequencies shown as opposed to percentage agreement to correct for any chance agreement (Robson, 2011). This found k = 0.86, indicating very good agreement.

5.3 RESULTS

5.3.1 DESCRIPTION OF SAMPLE

Participants were all fully qualified naval aircraft engineers and experienced at their specific employment. The engineers were grouped operatives and supervisors, according to the two main employment levels within a squadron, as opposed to the experience they have of a particular task. For example, operatives possess limited authorisations for aircraft maintenance that involves routine maintenance tasks such as aircraft flight servicing. A supervisor will have been an operative before promotion and is afforded wider authorisations for more demanding maintenance tasks such as in-depth aircraft fault diagnosis and additional functions such as planning aircraft maintenance and leading maintenance teams. The current study views each participant as equally experienced for their particular employment, and thus, the grouping of operative and supervisor simply reflects the general differences in how they are employed in a squadron.

The sample included both males (n = 45) and females (n = 3) from the target population, for which the low count of females in the sample is representative of the population. The sample consisted of 24 junior engineers (mean age = 27.6 years, SD = 4.3, range 21–39) and 24 senior engineers (mean age = 36.6 years, SD = 6.9, range 23–57). Trainees were not included as they are at an earlier stage of learning and thus still developing their maintenance skills (Fitts & Posner, 1967). As such, trainees are not authorised to conduct aircraft maintenance without 100% supervision, and additional safety checks are put in place to carefully monitor their work. Only one candidate chose not to participate based upon the belief that he had not experienced the phenomenon, and eight questionnaires were discarded after an initial read-through due to illegible handwriting or because the responses were not examples of post-task I-LED.

5.3.2 Analysis of Responses

5.3.2.1 Maintenance Task and Error Event

Questions 1–6 in Table 5.1 captured contextual data preceding the error event to identify potential I-LED influences. Responses to Question 1 all described familiar maintenance tasks that were routine and well practised, for example, aircraft servicing, component changes, completion of aircraft documentation, logistics and refuelling. Question 1 also showed 79.2% (n = 38) of maintenance tasks to be complex tasks, complex tasks being an aircraft engine change or flight servicing an aircraft (defined by naval aircraft maintenance policy). The remainder 20.8% (n = 10) were simple tasks such as checking a toolbox, securing an aircraft panel or signing for completed work in an aircraft maintenance document. Question 2 reported most engineers 60.4% (n = 29) were tasked with more than one maintenance activity at the time the error event occurred, having identified the maintenance task mostly through a process followed by a verbal brief then self-identified: 56.3% (n = 27), 41.7% (n = 20) and 2% (n = 1), respectively. Question 5 reported there was a delay starting 52% (n = 25) of the maintenance tasks that were later found in error, with 48% (n = 23) started immediately. The data from Question 5 did not provide sufficient detail to calculate the length of the delays.

Combined responses from Questions 2 and 7 are shown in Table 5.2, which highlights the location where a maintenance task was identified and where the actual error event occurred. For 42% (n = 20) of maintenance tasks, the maintenance requirement was identified whilst in the Aircraft Maintenance Coordination Office (AMCO), 37% (n = 18) in the hangar and 21% (n = 10) in other environments, including: maintenance offices; crew room; or whilst physically on an aircraft. Participants reported 52% (n = 25) of error events occurred in the aircraft hangar, followed by 21% (n = 10) on the line, 12% (n = 6) in the AMCO and 15% (n = 7) were other work locations such as a tool store or workshop. Responses to Question 7 also allowed Table 5.3 to be constructed, which shows each error event categorised according to Reason's (1990) GEMS error categorisation, where 69% (n = 33) of maintenance error events were categorised as a lapse with mistakes accounting for 27% (n = 13) and slips 4% (n = 2). Latent detection of execution errors (slips and lapses) represented 73% (n = 35) of the sample with planning errors (mistakes) accounting for 27% (n = 13). The results gave an equal overall count of error categories between junior and senior engineers, although senior engineers did not report suffering any slips.

TABLE 5.2
Locations Associated with Task Identified and the Error Event

Physical Location	Task Identified	Error Event
AMCO	20 (42%)	6 (12%)
Hangar	18 (37%)	25 (52%)
Line (Ramp)	4 (8%)	10 (21%)
Other	6 (13%)	7 (15%)

TABLE 5.3
Error Events Categorised According to GEMS

Error Category	Error Type	Count	Example
Execution error	Lapse	33 (69%)	Forgot to ensure wiring loom secured
(n = 35)	Slip	2 (4%)	Engine oil dipstick not seated correctly
Planning error	Mistake	13 (27%)	Wrong component fitted
(n = 13)			

Source: Reason, J. 1990. *Human Error.* Cambridge: Cambridge University Press.

5.3.2.2   Error Detection Themes

Error detection themes that emerged from the responses to Question 9 are recorded in Appendix C and summarised in Table 5.4. Thematic analysis revealed five main themes associated with different environments, along with a count of reported occasions. Themes for 'self-doubt/suspicion' and 'task-related cue' each account for 25% (n = 12) of the sample followed by 'error came to mind' 23% (n = 11), 'reflection/review' 21% (n = 10) and 'discussing work' 6% (n = 3). Table 5.4 also provides example narratives for each theme based upon whether the I-LED occurred At Work or Not at Work. At Work is argued to mean the same or similar working environment to that which the error occurred and Not at Work being other physical environments unrelated to the workplace, that is at home or driving a car. Thus, reading from Table 5.4, 62.5% (n = 30) of I-LED events occurred when At Work and 37.5% (n = 18) when Not at Work. To illustrate this delineation further, Figure 5.2 shows themes against physical environment and experience.

The five I-LED themes shown in Table 5.4 were grouped as Intentional Review (IR), where the engineer consciously and deliberately reviewed past events before detecting the latent, and Unintentional Review (UR), where the engineer did not instigate a conscious and deliberate attempt to review past events prior to detecting the latent error condition. Here, it is argued IR accounts for the theme 'reflection/ review', and UR accounts for the more spontaneous themes of 'self-doubt/suspicion', 'discussing work', 'came to mind' and 'task-related cue'. Table 5.5 highlights this grouping against different environments associated with employment and specific locations. The UR group accounted for 75% (n = 36) of all error detections where 50% (n = 24) occurred At Work, and 25% (n = 12) occurred when Not at Work. The IR group accounted for 25% (n = 12) where 12.5% (n = 6) occurred At Work, and 12.5% (n = 6) occurred whilst Not at Work. I-LED events were concentrated around the AMCO and the aircraft hangar, accounting for 33% (n = 16) of the specific locations At Work. Not at Work error detections were move evenly spread across the specific locations shown in Table 5.5.

Data in Table 5.5 allowed a 2×2 contingency table to be constructed using categories for employment level and environment, null hypothesis being I-LED is equally likely to occur for each employment level for either physical environment in which the detection occurred. Chi-square test showed no significant association between employment and the environment in which the past error was detected

TABLE 5.4
Detection Themes Associated with Environment

Detection Theme	At Work (n = 30)		Not at Work (n = 18)		Totals
	Frequency	Example Narrative	Frequency	Example Narrative	
Reflection/ review	5	'As I walk back to hangar...I visualised previous job and remembered seeing leads in breach position'.	5	'Driving home thinking about the day's work'.	10 (21%)
Task-related cue	11	'I went to return the tools and saw the spanner I used...then I realised I'd forgotten to tighten the bolts'.	1	'On return home that night I was watching TV...news showed Sea King [helicopter]. This made me think...'	12 (25%)
Error came to mind	7	'In the line office prior to next task...looking out of window and remembered that I hadn't removed blanks'.	4	'At lunch remembered putting a rag at the back of the engine'.	11 (23%)
Self-doubt/ suspicion	6	'Disbelief in my work so went to check'.	6	'At home watching TV at night I felt uneasy about something'.	12 (25%)
Discussing work	1	'Talking over paperwork, it occurred to me that I may not have tightened the screws'.	2	'In bar...was chatting with colleague... realised I had omitted check on cannon [gun]'.	3 (6%)

($\chi^2_{(1)} = 1.42$, p = ns), although 71% (n = 17) of operatives detected their past error At Work. A 2×2 contingency table was also constructed using the categories for review group and environment shown in Table 5.5, null hypothesis being IR and UR are equally likely to occur in each physical environment in which the detection occurred. Due to low frequencies, a Fisher's exact test was carried out, which showed no significant association between categories (P = 0.325, p = ns). Notably, 67% (n = 24) of URs take place At Work. Chi-squared test was also carried out on a 2×2 contingency table constructed from Table 5.5 using the categories for review group and employment, null hypothesis being IR and UR are equally likely to occur between operatives and supervisors. The test showed no significant association between categories (P = 0.37, p = ns). From Question 12 in Table 5.1, 75% (n = 33 of 44) of participants reported they had experienced a false detection, examples being similar to the routine maintenance examples already highlighted. The thematic analysis of the responses to Question 13 are recorded in Appendix D and summarised in Table 5.6, which places 20 themes determined from the narratives in ranked order of frequency.

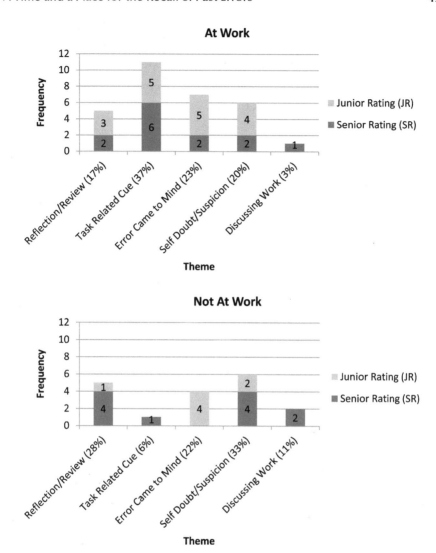

FIGURE 5.2 Frequency of themes related to environment and employment.

5.3.2.3 Timing Factors

Responses to Question 6, 8 and 11 allowed the following timing factors to be determined. Most maintenance tasks were identified and carried out in error during the hours of 0800–2000, which correlates with the normal working day when not deployed away from the unit (i.e., not at sea embarked in a ship). Four tasks were outside of the normal working day, and five participants did not report any timing data. Most narratives gave non-specific timings; thus, data were ranked according to the mean of the timings given by each participant. For example, if 1–1.5 hours were reported, then 75 mins was taken as the time for analysis. This process showed an outlier of 12,960 mins, which was sifted to give a total of 42 reports that were

TABLE 5.5
Environment Associated with Review Group and Specific Location

Environment	Review Group	Location	Operative	Supervisor	Totals
At work	IR	AMCO	1	0	6
		Hangar	1	1	
		Issue centre	1	0	
		Crew room	1	0	
		Walking	1	0	
	UR	AMCO	4	3	24
		Hangar	3	3	
		Line	3	1	
		Maintenance office	1	2	
		In aircraft	0	1	
		Walking	2	0	
		Onboard ship	0	1	
Not at work	IR	Bed	0	1	6
		At home	0	1	
		Showering	0	1	
		Driving a car	1	2	
	UR	At home	0	3	12
		Bar	0	1	
		Cabin (bedroom)	2	0	
		Driving a car	0	1	
		Mess	3	0	
		Walking	1	1	

analysed for timing factors. The time (T) between identifying a task (t) and the error event (e) was recorded as T(t-e), and the time between the error (e) and the latent detection (d) was recorded as T(e-d). From the timing data reported in the 42 responses analysed, the overall mean for T(t-e) was found to be 206 and 201 mins for T(e-d), and 60% (n = 26) of past errors were detected within an hour or less of occurring, whilst 70% (n = 30) were detected within two hours or less. To test for significance, Spearman's Rho was calculated for the null hypothesis that there is no correlation between T(t-e) and T(e-d). A moderate positive correlation was found ($r_s = 0.66$, $p < 0.01$). This correlation is shown in Figure 5.3 with further timing data reported in Table 5.7 to show employment and themes against mean times.

5.4 DISCUSSION

5.4.1 MAINTENANCE TASK

All participants who completed a questionnaire were able to recount personal examples of post-task I-LED, which aligns with widely accepted views that human performance variability and latent error conditions are common, suggesting the I-LED phenomenon is also prevalent (Reason, 1990; Hollnagel, 1993; Maurino et al., 1995;

TABLE 5.6

Themes Derived From Question 13 in Ranked Order of Frequency

Ranked Order	Ranked Frequency of Themes (n = 55)
1st	Check work (n = 13)
2nd	Use checklist/process (n = 5)
	Avoid task pressure (n = 5)
3rd	Avoid interruptions/distractions (n = 4)
	Rest/break needed between tasks (n = 4)
	Experience/skill needed (n = 4)
4th	Training must be fit for purpose (n = 3)
	Error awareness (n = 3)
5th	Avoid tasks in parallel (n = 2)
	Avoid rushing (n = 2)
6th	Individual responsibility (n = 1)
	Employer responsibility (n = 1)
	Reward/incentives (n = 1)
	Processes must be fit for purpose (n = 1)
	Emphasise task importance (n = 1)
	Avoid delays (n = 1)
	Just culture (n = 1)
	Use warnings (n = 1)
	Influence of human performance (n = 1)
	Influence of personality (n = 1)

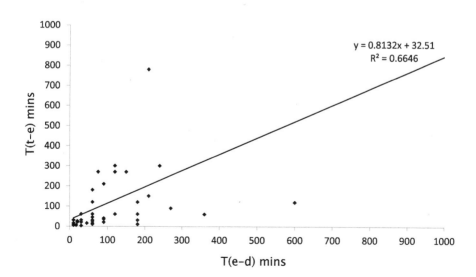

FIGURE 5.3   Correlation of timing data, T(t-e) and T(e-d).

TABLE 5.7

Mean Detection Times for Employment and Detection Theme

Employment/Detection Theme	Mean Time T(e-d)
Operative	83 mins
Supervisor	297 mins
Reflection/review	195 mins
Task-related cue	343 mins
Error came to mind	61 mins
Self-doubt/suspicion	63 mins
Discussing work	310 mins

Perrow, 1999; Flin et al., 2008; Woods et al., 2010). Participants described routine maintenance tasks that were well practised and thus habitual in nature, which is characteristic of highly developed schemata (Bartlett, 1932). Most responses also described carrying out several complex maintenance tasks such as an engine change or main gearbox replacement, although all the error events related to simple tasks, for example, forgetting to sign the aircraft documentation (simple task) as part of the overall task of aircraft flight servicing (complex task). This suggests I-LED is more likely to be associated with the detection of simple tasks (i.e., no one reported that they later recalled they had forgotten to carry out an entire flight servicing task). Since schema theory offers that only high-level schemata need conscious control for subordinate schema to trigger autonomously, the current data appear to support this position, as the simple tasks described in the study were largely autonomous and triggered as a result of a higher-level task requirement (Norman, 1981; Mandler, 1985; Cohen et al., 1986; Reason, 1990; Baddeley, 1997). In terms of PM theory, maintenance tasks were identified via a verbal brief or as part of a process, and since no participant reported using an external memory store such as written notes or an electronic aid, it is likely that the engineer formed the intent to carry out the maintenance in the mind, thereby creating an internal 'to-do' list for subsequent PCM processing (Sellen et al., 1996).

5.4.2 ERROR EVENT

Participants were asked to describe I-LEDs (i.e., they were not to recount errors detected post task via an existing LED safety strategy such as process or supervisor checks). Using the GEMS categorisation of error (Reason, 1990), results showed the category and ratio of errors self-detected by the engineers to be broadly representative of wider maintenance error studies (Reason, 1990; Graeber & Marx, 1993; Latorella & Prabhu, 2000), which provides confidence that the current findings can be read across wider contexts. Additionally, the error category did not appear to be affected by age or employment, provided the engineer was experienced for the required task.

Lapses were found to be the largest type of maintenance error detected via the I-LED phenomenon. Overall, there were recollections of a greater number of execution errors (slips and lapses) than planning errors (mistakes). The finding for execution errors included a low count of slips. This should be expected according to Sellen (1994), who argued this error type is likely to be detected proximal to the error event since the error is often evident immediately (i.e., dropping a tool). Blavier et al. (2005) found from their study of proximal errors that execution errors increase with complexity. Complexity is argued to be synonymous with aircraft maintenance environment, and thus, this may account for the high number of execution errors found in the study since, extrapolating Blavier et al.'s (2005) finding, a proportional number could have become latent error conditions that were detected via I-LED. Further, the dominant number of lapses aligns with Reason (2008), who found lapses to be the most frequent error type in this category. Woods (1984) and Kontogiannis and Malakis (2009) found that the proximal detection of execution errors is higher than planning errors. From a PCM perspective, planning errors are also less likely than execution errors to be detected proximal to the error event due to the implicit lack of awareness of a genotype/phenotype mismatch as important external cues are not recognised during schema processing (Norman, 1981; Plant & Stanton, 2013b). This is characteristic of someone who possesses an incorrect perception of the external environment, leading to insufficient Situational Awareness (SA) for a particular task (Cohen et al., 1986; Plant & Stanton, 2013b). Thus, this potential lack of awareness of planning errors that occur proximal to task execution may explain the number of mistakes that were later detected via the I-LED phenomenon.

Wilkinson et al. (2011) found experts have a special ability to detect errors, which is exhibited as an enhanced 'capacity' to detect important cues present in the external environment. This is arguably an attribute associated with the engineers in the current study, since each participant was experienced for their particular employment and therefore more likely than trainees to benefit from an enhanced capacity to detect important cues distributed across the sociotechnical system. Consequently, this enhanced capacity also enhances or 'strengthens' schema processing to detect and respond correctly to important cues, which assists with the regain of SA (Brewer et al., 2010; Plant & Stanton, 2013a; Rafferty et al., 2013). Thus, when a regain of SA does occur in the form of I-LED, it is argued that a higher number of lapses should be detected over mistakes or slips, as a higher proportion is not detected proximal to the error event, and therefore, proportionally, more become latent error conditions.

5.4.3 Error Detection Themes

There was no significant difference between the detection themes except for the low frequency of participants reporting the 'discussing work' theme. Results showed post-task I-LED occurred more often At Work than Not at Work. Schema and distributed cognition theories could explain this finding for an operator immersed in a rich environment of distributed memory cues (Bartlett, 1932; Norman, 1981; Hutchins, 1995). For example, Grundgeiger et al. (2014) reported visual cues in the workplace increased occasions of PM recall compared to no cues. The narrative 'saw the spanner' highlighted in Table 5.4 exemplifies the impact of visual cues and is likely to account for the dominant number of 'task-related cues' At Work (see Figure 5.1, At Work). However, for other At Work detection themes highlighted in

Table 5.3, the link to distributed cognition is less clear, for example, engineers who reported visualising a previous job, looking out the window or possessing a general disbelief in their work. These occasions are considered when discussing conceptual groups UR and IR.

When Not at Work, and therefore remote from the maintenance context, 18 examples of I-LED were reported. Referring to Figure 5.1 (see Not at Work), expectantly 'task-related cues' contributed the least to I-LED (n = 1), with 'self-doubt/suspicion' the dominant category. The example in Table 5.4 reports feeling 'uneasy about something' when watching TV at home. Distributed situational awareness (DSA) theory describes how SA is dependent on distributed cognition within the entire sociotechnical system, forming external memory structures within the physical world such as the aircraft maintenance environment (Hutchins, 2001; Stanton et al., 2015). For an I-LED event to occur whilst watching TV, for example, it is argued the cognitive capability of the experienced engineer must be distributed across unrelated environments, and thus, DSA theory appears to span other sociotechnical systems, suggesting the entire sociotechnical system is represented by interlinked sociotechnical contexts (At Work and Not at Work). For example, a main rotor gearbox change is a lengthy procedure requiring the hierarchical activation of multiple schemata. The experienced engineer will possess sufficient skill-rule-knowledge-based behaviour (Rasmussen, 1982) to carry out the task they are trained to do, for which schema activation is supported by distributed memory stores (cues) in the maintenance environment such as written procedures, aircraft documentation, the physical aircraft, tool procedures and so on. Post-task completion, data suggest these typical cues or external memory stores remain available for latent review of the 'to-do' list or internal markers. This facilitates a persistent opportunity for schema housekeeping, and results appear to show this leads to I-LED both At Work and Not at Work. Smith et al. (2007) found PM retrieval also requires continuous monitoring for relevant cues to achieve task success, whilst Grundgeiger et al. (2014) reported the phenomenon of automatic retrieval, for which it is argued SAS monitoring (Norman & Shallice, 1986) must exist post-task completion for I-LED to occur; a function of multi-process behaviour leading to the review of past activity.

The results showed 75% (n = 36) of I-LEDs were grouped as UR; 24 I-LEDs occurred At Work, and 12 occurred when Not at Work. For the latter, four I-LEDs occurred when resting and eight when engaged in unrelated activities such as watching TV, walking, showering, driving and so on. Arguably, all 12 UR examples occurred when engaged in well-learnt and non-complex activities, which is indicative of largely autonomous behaviour requiring unfocused attention or limited focal processing (Fitts & Posner, 1967; Brewer et al., 2010; Rasmussen & Berntsen, 2011). Involuntary Memory (IVM) research reports the spontaneous recall of a memory, which bypasses any conscious attempt to recall a specific memory and at some time after the PM intention was needed (Haque & Conway, 2001; Blavier et al., 2005). Rasmussen and Berntsen (2011) later argued that IVM is mostly associative and context dependent and usually cued by some aspect of the physical environment,, which is most likely to occur when attention is unfocussed and thus requiring minimal execution control. This aligns with schema processing (Neisser, 1976), and whilst their work considered autobiographical IVM and did not take a systems approach, its argued similarity to the current research can be drawn, since developed schemata supports autonomous behaviour that characterises the experienced engineer and the engineer's unconscious ability to exploit DSA (Hutchins, 1995; Stanton et al., 2015). Sellen (1994) found that PM recall was more often due to contextual factors, as opposed to spontaneous or controlled recalls, and that creating associations improved retrieval of relevant information. This aligns with Bartlett (1932) who argued 'literal' recall is less likely than recall as a result of reconstruction or interpretation of past events. Although Haque and Conway's (2001) spontaneity aspect is at odds with Sellen, it is thought more likely to be the result of the autonomy of schema housekeeping (Saward & Stanton, 2015a), as data for URs suggest the individual did not have control over schema processes, whilst the argument for unfocussed attention appears accurate as all examples of I-LED occurred during periods of autonomous behaviour. This position also receives support from Plant and Stanton (2013b) who argued schemata are selected in a hierarchical pattern determined by the strength of the target activity or task (level of focal processing). Further, studies have shown that the conflict between ongoing task monitoring and continuous monitoring for cues leads to interruptions or interference, which decreases operator performance (Norman & Bobrow, 1975; Einstein et al., 1997; Marsh et al., 1998). The impact of this divided attention can be influenced by the level of focal processing needed for a task (Brewer et al., 2010) and thus offers some explanation why periods of unfocussed attention provide SAS the opportunity for schema housekeeping to take place; this catch-up period potentially influences I-LED.

Thus, wider research appears to agree that UR (and IR) requires post-task reconstruction of the sociotechnical context to detect errors, which can occur during periods of unfocussed attention. Distributed cognition is also believed to span unrelated environments and extend into the mind, for which the experienced engineer can visualise the relevant sociotechnical system and reconstruct past events for I-LED to occur. Grundgeiger et al. (2014) argued recall can occur at work and in everyday unrelated environments, and, arguably, this describes further the entire sociotechnical system. Since correct SA is considered to rely on schema processing of relevant cues as part of the PCM cycle, it is also argued SAS monitoring is persistent and able to occur in unrelated environments for past incorrect perceptions (schema mismatch) to be detected latently. This supports the concept of Latent Error Searching (LES) described in Chapter 2 within the detection themes self-doubt/suspicion and reflection/ review that appeared to require reconstruction of past maintenance tasks to search for error. Thus, At Work, whilst I-LED is most successful when post-task schema housekeeping takes place in the same or similar environment to that which the error occurred, significantly, the engineer appears to also possess the cognitive capability for remote access to these cues when Not at Work by internally contextualising past events. This extends PCM theory, which already reports to be a reliable method to account for human sociotechnical interaction within a given environment but now appears more widely distributed over interlinked sociotechnical contexts, coupled with persistent SAS monitoring (Plant & Stanton, 2013b). Notably, this is a characteristic of distributed cognition where correct operation in a sociotechnical system relies on internal and external structures or memory stores (Hutchins, 2001). Interpreting this further, by the very nature of being an experienced aircraft engineer, the engineer is very familiar and intimate with these schema cues, and thus, it is offered that elements of the sociotechnical system have been 'absorbed' and are available in the mind if needed. This may offer an explanation as to why many past errors were detected when remote from the working environment and also why the I-LED event appeared spontaneous to the engineer (supporting Haque & Conway, 2001). The net effect is that distributed cognition appears to extend beyond specific sociotechnical environments where an experienced engineer's ability to reconstruct events (and therefore identify system cues) in any environment is sufficient to trigger schema housekeeping, which offers the opportunity for I-LED as the product of collaborative multi-processes. This also highlights an enduring linking of the mind back to the external environment relevant at the time the error occurred (Hutchins, 2001; Stanton et al., 2015). If system cues trigger schemata, and these cues remain proximal or linked remotely in the mind, then it appears the distributed cognition concept extends to I-LED and is effective across everyday sociotechnical contexts.

Participants perceived checking their work for missed errors using a checklist/ process to be most effect I-LED intervention according to the themes determined from Question 13. This is an interesting finding as it arguably reinforces Reason's (1997) heroic concept, since the aircraft engineers in this study are keen to exploit ways that they can locally manage their detection and recovery from past errors (as opposed to relying on an independent checker). Aircraft engineers are also highly process-driven, which may drive their desire for a checklist/process, although this in itself may indicate a lack of safety resilience in existing control measures such as maintenance processes or training. The concept of resilience and the integration of targeted LED interventions to achieve total safety are discussed in Chapter 7. The participants also suggested that I-LED events might be enhanced further if the check is conducted alone (avoiding interruptions/distractions) during a break/rest between tasks, thereby avoiding task pressure. Arguably, the themes captured in Table 5.6 could be indicative of weaknesses within the sociotechnical system associated with maintenance processes, which may benefit from targeted practicable interventions using system generated cues. Notably, most LED suggestions provided by the participants are arguably organisational control measures enacted and managed locally by individuals (and teams) to maintain safety equilibrium through safe behaviours. This affirms the need from a human-centred approach to system safety based on understanding of typical human error effects, which was discussed in Chapter 3.

The potential for false error detections has also been reported and is argued to be a consequence of the largely autonomous nature of schema activation and persistent enquiry of the PCM through ongoing SAS monitoring (Norman, 1981; Mandler, 1985; Cohen et al., 1986; Baddeley, 1997; Saward & Stanton, 2015a). Results showed 75% of participants reported they had experienced false alarms, which indicates multi-process I-LED behaviour can lead to false detections. Data fidelity did not allow further exploration of this observation, and thus, it is not known why distributed memory cues (either external or internally contextualised) triggered a false detection. Arguable though, this phenomenon may indicate the PCM is not infallible and perhaps further supports the earlier view that schema housekeeping is indiscriminate. Thus, this may explain I-LED occurrences that seemed to be detected by chance (or false detection) under the themes of 'came to mind' and 'self-doubt/suspicion'.

5.4.4 Timing Factors

Data indicated significant correlation between T(t-e) and T(e-d) where 70% (n = 30) of detections occurred within two hours of the error event and 60% (n = 26) within an hour. This is an intriguing result as it suggests the PCM cycle possesses a particular frequency, for which a 'golden time window' of opportunity may exist for I-LED post-task completion. This appears to support the Patel et al. (2011) finding that most errors are self-detected and recovered during the working period, which could provide an explanation for the supposedly spontaneous nature of detections reported by the engineers where later events appear to trigger I-LED.

The frequency of schema housekeeping may also be dependent on detection theme. Data shown in Table 5.6 revealed the lowest average I-LED times were for errors that came to mind and self-doubt/suspicion, both UR themes. This may be indicative of different influences on I-LED, that is the dependence on the autonomous PCM cycle to detect mismatches within a potential golden time window. As discussed, most detections occurred whilst At Work, and thus, this finding is probably not surprising. However, whilst most task-related detections occurred At Work, the average time for this category was the highest. This is surprising as it was anticipated the At Work environment would give rise to rapid I-LED via abundant task-related cues; thus, it may infer strong habit intrusion amongst engineers who are de-sensitised (and therefore low focal processing) due to the strength of these distributed memory cues. Of note, no significant difference between operative and supervisor across detection themes was found, provided they are experienced for the task required of them. The mean detection times in Table 5.6 show operatives did detect more errors at work and more quickly, but this is considered to be due to their employment as junior engineers who spend the majority of their working day in the hangar or line, which is rich in task-related cues. Whilst this research is at an exploratory stage, the very existence of I-LED demonstrates time is a factor and for which it is argued system cues must be persistent for post-task schema housekeeping; otherwise, I-LED is not possible.

5.5 SUMMARY

Human error is inevitable and a daily occurrence, which provides impetus for the discovery of mitigation strategies through the safety research described in this book. Understanding post-task I-LED may provide partial mitigation to achieve error resilience, yet it appears to be an under explored safety field. Objective 2 looked to progress understanding of the nature and extent of the I-LED phenomenon by observing naval aircraft engineers in the workplace, against which the current study hypothesised time, location and other system cues influence I-LED as the basis for an initial exploration using a multi-process framework applied to a questionnaire administered during group sessions. Data from the aircraft engineers confirmed the existence of I-LED, which appears to be prevalent within the naval aircraft maintenance environment. Having formed intent to carry out a task (or tasks) and therefore created a 'to-do' list in the mind, engineers mostly suffered lapses involving simple tasks that were later detected. However, lapses may have dominated as a consequence of the reduced ability of genotype/phenotype processing to detect mistakes due to the implicit lack of awareness of importance system cues.

Latent error detections occurred more often At Work than Not at Work; thus, I-LED was found to be more successful when post-task schema housekeeping most likely takes place in a physical environment that is the same, or similar to, the physical environment in which the actual error occurred. DSA recognises that the strength and distribution of system cues are important for schema triggering; thus, it should be expected that the 'task-related' theme dominated At Work due to a high concentration of work-related cues in the aircraft maintenance environment. The 'self-doubt/suspicion' theme appeared dominant when Not at Work despite detection themes for 'reflection/review' and 'error came to mind' being conceptually similar.

It is argued collaborative multi-processes that include theories on PM, SAS monitoring and schemata are effective for observing the I-LED phenomenon, which can manifest in other physical environments unrelated to that which the error occurred. Also, the very existence of I-LED indicates important cues remain available for post-task schema housekeeping to detect past errors, for which there may be golden window for successful I-LED. Here, it is thought the entire STS may be represented by interlinked sociotechnical contexts from which the air engineer can remotely access relevant cues by visualising/reconstructing past tasks in the mind. Combined with the argument that schema housekeeping is largely autonomous and persistent, this may also explain why the air engineers reported Not at Work themes widely. This extends distributed cognition thinking beyond the need to remain proximal to the error event or physically immersed in the same system contexts to where the error actually occurred. Further, the argument that post-task schema housekeeping may be indiscriminate provides an explanation for occasions of false alarms, or conversely, chance detections.

It is also argued time, location and other systems cues trigger post-task I-LED. Using a sociotechnical approach combined with multi-process theory advances current systems thinking, whilst unlocking the role of PCM and distributed cognition in I-LED may lead to systemic interventions aimed at improving safety resilience. Further research is needed to mature the exploratory nature of the current study: to understand how the entire system of interlinked sociotechnical contexts contribute to I-LED; how the PCM links to the world in which the error occurred for the visualisation/reconstruction of past events; and consequently, to develop practicable interventions to enhance post-task I-LED within a potential golden time window of opportunity as close to the creation of the latent error condition as possible. Since the concept of human error has broad applicability, it is anticipated current research will be beneficial to the wider community interested in safety resilience using a systems perspective to minimise the consequences arising from undetected error.

This exploratory study confirmed the presence of the I-LED phenomenon in a cohort of naval aircraft engineers, which addresses Objective 2 described in Chapter 1. The following chapter expands on the findings of the current study to addresses Objective 3 by observing a further cohort of naval air engineers in operating squadrons using a diary study. The main aim of the following study reported in Chapter 6 is to identify practicable I-LED interventions that can be used during routine normal operations to help control human error effects and, therefore, enhance safety resilience.

6 A Golden Two Hours for I-LED

6.1 INTRODUCTION

As discussed in previous chapters, undetected error in any safety critical system becomes a latent error condition that can contribute to a future safety failure; thus, the detection of past errors is an essential element of resilience safety management (Rasmussen & Pedersen, 1984; Reason, 1997; Shorrock & Kirwan, 2002; Wiegmann & Shappell, 2003; Flin et al., 2008; Aini & Fakhru'l-Razi, 2013). Due to the apparent paucity of I-LED research, Chapter 2 reviewed literature for transferrable theories. This led to a multi-process approach to I-LED research that combines theories on Prospective Memory (PM), Supervisory Attentional System (SAS) monitoring and schemata. The PM element refers to the creation of intent to carry out an action and the SAS for monitoring of the schema-action-world cycle, which is characterised by the Perceptual Cycle Model (PCM). Schema theory, embedded within the PCM, highlights the human interaction with the Sociotechnical System (STS) via the bottom-up (BU) processing of external sensory data against top-down (TD) knowledge of the world (schemata) within a perceptual cycle (Neisser, 1976; Cohen et al., 1986; Plant & Stanton, 2013a). This forms the transactional relationship between a schema-action-world cycle and system cues in the external world that trigger intended actions (Neisser, 1976; Norman, 1981; Mandler, 1985; Stanton et al., 2009b; Plant & Stanton, 2013a). Studies on PM indicate that intentions (schema selection) are 'loaded' into memory to act upon later, which generates a 'to-do list' or internal marker (Sellen et al., 1996; Marsh et al., 1998; Van den Berg et al., 2004). The SAS described by Norman and Shallice (1986) is thought to be the attentional mechanism to continually monitor the external environment for external cues and monitor the perceptual cycle for correct execution of intent on the 'to-do' list (Norman, 1981; Norman & Shallice, 1986; Smith, 2003; Einstein & McDaniel, 2005; Saward & Stanton, 2015b). The SAS is also argued to regulate schema housekeeping, which is a term used to simply highlight the function of monitoring the perceptual cycle to confirm an action is completed as well as collecting feedback from the action to facilitate learning and the acquiring of experience (Saward & Stanton, 2015b). Developed schemata that form internal memory structures, which can be accessed to respond to a particular task, are known as genotype schemata, whilst phenotype schemata refer to the actual response when executing a task (Neisser, 1976; Reason, 1990). Thus, schema housekeeping is thought to highlight the cyclic update of genotype schema (for schema learning) by reviewing previous schema-action-world information and thereby providing the opportunity for the detection of past errors (Saward & Stanton, 2015b).

PM research has also found that the successful recall of intentions is cue dependent and can be triggered automatically by external cues present in environmental contexts (Tulving, 1983; Guynn et al., 1998; Bargh & Chartrand, 1999; Einstein & McDaniel, 2005). Particularly, easily recognised cues in the external environment, being mostly visual or auditory, are effective triggers of internal markers, where written word cues have been found to be more likely to trigger recall than picture cues (Kvavilashvili & Mandler, 2004; Mazzoni et al., 2014).

Sellen (1994) offered that an operator could fail to detect an error because the success of the action could have been imperceptible, which indicates the schema-action-world cycle is dependent on cue recognition (e.g., where an oil filler cap was not quite seated correctly or a similar, but the wrong lubricant was used on a component). Autonomous schema housekeeping may later highlight the genotype/phenotype mismatch. This accounts for why past errors can appear to come to mind spontaneously (Reason, 1990; Einstein & McDaniel, 2005; Stanton et al., 2009b). If cue information was imperceptible at the time an activity was carried out, then differences in the activity or location associated with an I-LED event needs consideration. Findings from Chapter 5 highlighted the link between I-LED and cognition distributed across different sociotechnical environments (Hutchins, 2001; Stanton et al., 2015; Grundgeiger et al., 2014). Here, I-LED was argued to be most successful when post-task schema housekeeping takes place immersed in the same environment to that which the error occurred, although latent error conditions arising in the workplace were also detected similar or unrelated surroundings. This extended task-related cue recognition across a range of unrelated sociotechnical environments where different environments could be accessed for cue information by internally visualising/reconstructing past activity, for example, when at home, driving a car, walking, showering and so on. Further, and by extending transferable theories on memory, I-LED is thought more likely to occur when alone (not interacting with others) and mostly during periods of unfocussed attention such as inactivity, day dreaming or engaged in largely autonomous activities that do not require high levels of concentration on the task in hand (Kvavilashvili & Mandler, 2004; Smallwood & Schooler, 2006; Rasmussen & Berntsen, 2011).

PCM theory describes a schema-action-world cycle (Plant & Stanton, 2013a); therefore, time must be associated with the frequency of the cycle. It was argued in Chapter 5 that the perceptual cycle persistently reviews past task performance through schema housekeeping, for which distributed cues must remain available across sociotechnical environments for I-LED events. This could facilitate a 'golden time window' for I-LED to occur. Initial findings indicated most latent error conditions were detected within a time window of two hours of occurring, which receives some concurrence from Patel et al. (2011) who highlighted 75% of errors committed by expert clinical staff were detected and recovered within 10 hours whilst at work. It was also argued that this persistence is largely autonomous, leading to the unintentional review of a past task that perhaps accounts for seemingly spontaneously chance detections, although the intentional review of a past task is expected to be more successful than the autonomous condition.

Using the multi-process approach to systems thinking described previously, the following study is designed to advance existing literature associated with I-LED via the real-world study of naval aircraft engineers. To determine whether this cohort exhibits normal cognitive behaviours, literature was reviewed for an appropriate instrument to employ. The Cognitive Failures Questionnaire (CFQ) in Appendix E scores an individual's propensity for everyday cognitive failures using 25 questions scored 0–100 against a five-item Likert coding (Broadbent et al., 1982). A high mean score (>51) indicates a propensity for cognitive failures (Broadbent et al. 1982). Whilst organisational safety performance has rightly moved away from focusing on individual human failings to a system-induced view of erroneous acts, knowledge of individual performance variability within a defined cohort is argued to remain important in anticipating the level of resilience that must be engineered into safe systems within the workplace (Reason, 2008; Woods et al., 2010; Reiman, 2011; Cornelissen et al., 2013). Wallace et al. (2002) administered the CFQ questionnaire to US Navy personnel, whilst Bridger et al. (2010) studied a large cohort of naval personnel in the Royal Navy. Both found these cohorts to exhibit normal performance variability representative of skilled workers, which is a relevant benchmark to inform the current study. Literature also indicates that those with a high CFQ score are more susceptible to performance variations leading to erroneous acts due to poor executive function, yet they are likely to have developed a personal coping strategy in the knowledge that they are prone to cognitive failures (Reason, 1990; Wallace et al., 2002; Mecacci & Righi, 2006; Day et al., 2012). Applying systems thinking, it is argued that this should be interpreted differently: that those with a high CFQ score are likely to be less receptive to external cues that trigger the necessary schema response, for which unreported behaviours may have been created that engage with the sociotechnical environment to engineer resilience.

Thus, the aim of the study described in this chapter is to understand how individuals engage with system cues for successful I-LED so that practicable interventions can be identified to make the system safer through enhanced resilience. This supports Objective 3 stated in Chapter 1, against which it is believed the exploratory findings from Chapter 5 will remain applicable in a real-world study and confirm that most I-LED events occur within two hours of the erroneous act, significantly, when alone during periods of unfocussed attention. Further, sensory data from familiar everyday cues present within the engineer's workplace are expected to facilitate successful I-LED through engagement with the perceptual cycle, which is a finding from the study conducted in Chapter 5. The hypothesis drives the requirement for real-world study to understand what promotes Safety II behaviour in the workplace in terms of I-LED, where it is believed further safety resilience is achievable through successful latent error detections. A diary study is selected as an effective method to capture everyday I-LED events over a protracted period in naturalistic environments and without biasing the data through intrusive observations (Reason, 1990; Cassell & Symon, 2004; Robson, 2011), whilst the CFQ is administered to simply affirm normal cognitive behaviours in naval air engineers by relating to research from wider populations.

6.2 METHODOLOGY

6.2.1 PARTICIPANTS

A convenience sample was conducted (Robson, 2011), which comprised representative numbers of operative and supervisor naval air engineers from the target population in RN helicopter squadrons. Aircraft engineers generally all train and operate to the same standards and practices; thus, significant differences do not exist between squadrons in terms of the working environment or employment, which need accounting for in the analysis of data. Six squadrons were available for the study, consisting of 695 engineers, of which 173 engineers participated (mean age = 29.99 years, sd = 6.81, range 18–48). This represents 25% of the population and includes both males (n = 164) and females (n = 9). Female participants accounted for 5.2% of the sample, which is representative of the population. As the low count of females is not statistically significant, no separate analysis of female responses could be conducted within the scope of the current study. Flanagan (1954) argued that the number of events was more important that number of participants, for which Twelker (2003) recommended no less than 50 events were needed for data to be meaningful. Thus, 173 participants were considered acceptable to yield sufficient events for analysis within the resources available for the study, although 60% attrition was anticipated due to participant dropout or unusable diary entries; thus, the minimum number of returned diaries was expected to be 70 and therefore sufficient for analysis.

6.2.2 DIARY DESIGN

A self-report diary was used to capture everyday I-LED events observed in the workplace, thereby avoiding intrusion but with adjacency and detail (Reason, 1990; Cassell & Symon, 2004; Robson, 2011). The diary in Appendix F was constructed according to Flanagan's (1954) Critical Incident Technique (CIT) where the term critical simply refers to a significant I-LED event reported by the participant. Neutrally worded questions were generated according to multi-process theories shown in Table 6.1. Intentionally, the diary was not designed against questions from the CFQ as the diary was designed to capture system factors. Free text descriptions of I-LED events were avoided since Schluter et al. (2008) found experienced nurses found it hard to describe their error behaviours. Questions were designed to give largely quantitative responses, as the exploratory study in Chapter 5 was qualitative. To help further ensure construct validity, diary methodologies were reviewed for good practice (Oppenheim, 1992; Sellen et al., 1996; Cassell & Symon, 2004; Johannessen & Berntsen, 2010; Mace et al., 2011; Robson, 2011).

6.2.3 PILOTING

A squadron not involved in the main study was approached for 10 aircraft engineers to practice administration and test the diary booklet. A small group of university research staff also tried the diary for general usability and question comprehension. Feedback from the pilot was provided via a follow-up session. Based on Wiegmann and Shappell's (2001) guide for an effective taxonomy, participants were asked to

TABLE 6.1
Diary Questions

Factor	Question	Response Options (Additional Comments in Brackets not Published in Diary)
PM	Q1. Please give a brief description of the error event.	General narrative (to understand context for I-I-LED event)
Time	Q2. At what time did the error event occur?	Time of day
PCM	Q3. What type of task was it?	Complex/Simple/Don't know (looking for task complexity)
Cue	Q4. What was the cue to do this task?	Event/Time/Both
PCM	Q5. What was the error type?	Slip/Lapse/Mistake/NK (according to Reason's (1990) GEMS)
Location	Q6. Where were you when the error occurred?	AMCO/Hangar/Line/Maintenance office/Issue centre/Storeroom/Aircraft/Workshop/Flight Deck/Other (At Work locations)
Time	Q7. At what time did you recall the error (post task completion)?	Time of day (to calculate time between the error occurring and detection)
Location	Q8. Where were you when you recalled the error?	AMCO/Hangar/Line/Maintenance office/Crew room/Issue centre/Storeroom/Aircraft/Workshop/ Flight Deck/Home or Mess/Bed/Vehicle/Gym/Other (At Work and Not at Work locations)
SAS	Q9. What were you doing when you recalled the error?	Planning/preparing maintenance activity/Conducting similar maintenance activity/Conducting dissimilar maintenance activity/Walking/Driving a vehicle/Exercising (e.g., cycling, jogging/ Showering/Eating/Socialising (e.g., in a bar)/General work-related discussion/Daydreaming/ Resting/Entertainment (i.e., reading, TV, Internet, etc.)/Sleeping/Other
SAS	Q10. Did you intentionally review your past tasks/activities?	Yes/No
SAS	Q11. (If Q10 'yes') Was this part of your personal routine?	Yes/No
PCM	Q12. On checking your work, was the error:	Real/False alarm (looking for successful detection of a latent error)
Cue	Q13. Did anything in your immediate location appear to trigger the error recall?	Sound/Equipment/Document/Smell/Taste/General vista/Other (looking to identify system cues)
Cue	Q14. What were you thinking about at the time of the error recall?	Work-related thoughts/Non-work-related thoughts
SAS	Q15. Were you alone when the error was recalled?	Yes/No
PCM	Q16. The specific error was very clear to me.	Likert coding: Strongly Agree = 1, Strongly Disagree = 5 (Clarity of error)
PCM	Q17. I was very confident that my past task was in error.	Likert coding: Strongly Agree = 1, Strongly Disagree = 5 (Error confidence)
SAS	Q18. The error recall occurred when I was highly focused on the activity at Q9.	Likert coding: Strongly Agree = 1, Strongly Disagree = 5 (Task focus)

comment on: the comprehensiveness of the diary questions; whether the questions were sufficiently wide ranging and captured everything they wanted to record about their I-LED event; and how usable they found the diary booklet. The general readability of the Participant Information Sheet (PIS) shown in Appendix G was also assessed against the Flesch reading ease score and amended to achieve a score of 60.1 (standard readability). Based on piloting, changes were also made to the administration and diary booklet to remove repetition, typographical errors and ambiguity in some questions.

6.2.4   Data Collection Procedure

Approval for the study was received from local engineering management prior to participants receiving a standardised verbal brief, which included an explanation of each diary question. They were also provided with the participant information sheet in Appendix G. Naval aircraft engineers receive flight safety briefs and training on error types (refer to Figure 2.3, GEMS: Reason, 1990) as part of the UK Ministry of Defence (MoD) aviation error management system. However, the authors confirmed participant understanding of error types during the verbal brief, and instructions were printed in the diaries, which included examples. The CFQ, participant register and consent forms were then completed prior to issuing the diary booklet. Participants were asked to record each I-LED event as near to the occurrence as possible to counter memory decay effects. To avoid the completion of the diary causing an unsafe distraction, a notepad was included in the booklet for participants to make quick notes for later completion. The notepad also allowed participants to record any additional comments they wanted to record to avoid limiting any important data not considered in the design of the study. After two months, the lead author personally collected completed diaries to preserve anonymity from line managers.

6.3   RESULTS

6.3.1   Description of Sample

The percentage of engineers who returned their diaries was 37% (n = 64), which is close to the anticipated maximum of 40% determined from piloting. The mean age of those who returned their diary was 30.70 years (sd = 7.41, range 19–48); 38% (n = 24) of returned diaries were blank, as the participant had not experienced an I-LED event during the two months of the study. The 40 completed dairies contained 51 usable entries, after 13 entries were dismissed due to conflicting responses or the recorded error example was not an I-LED event. Overall, the minimum number of CIT events recommended by Twelker (2003) was achieved. A Kolmogorov-Smirnov test on age within the sample of 64 engineers who returned their diaries showed $D(64) = 0.13$, $p < 0.01$, indicating the distribution of sampled mean ages deviated from a normal distribution, positive skews towards a mode of 28. However, this is representative of the population in naval aircraft squadrons where there are approximately 2.5 times more operatives of a younger age than older supervisors.

6.3.2 ANALYSIS

Category variables from the diary questions were mapped against each other to construct a simple 87×87 matrix. The matrix was not used for analysis except to facilitate a targeted approach to data analysis. The matrix provided a general indication that I-LED events were particularly associated with simple event-based tasks involving lapses. These events were mostly detected accurately (few false alarms) in the workplace without the intentional review of a past task and whilst attending to an ongoing task working alone. Here, thinking about work and the presence of physical objects (cues) appear related. Thus, the following analysis focuses any these areas.

6.3.2.1 CFQ Scores

The mean CFQ score for the 64 engineers who returned their diaries was M = 38.00 (n = 64, sd = 10.77, se = 1.34), for which a Kolmogorov-Smirnov test showed $D(64) = 0.10$, p = n.s, indicating the distribution of sampled mean CFQ scores do not deviate significantly from a normal distribution. The mean CFQ score for air engineers who returned their diary but reported no I-LED events was M = 35.71 (n = 24, sd = 10.56), whilst M = 39.37 (n = 40, sd = 10.78) for those reporting a I-LED event. Whilst the mean CFQ score was slightly higher for those who reported an attentional failure, the mean is still low within normal range, and a t-test showed no significance between group means (t = 1.32, df = 62, p = n.s), although a small effect exists (r = 0.2). Participants were also asked whether they intentionally reviewed (IR) past tasks for errors or if recall appeared to be spontaneous, and therefore an unintentional review (UR) occurred. Those with a high mean CFQ score (≥ 51) reported slightly more UR (57%, n = 8) than IR (43%, n = 6), and those with a low mean CFQ score reported more URs (70%, n = 26) than IRs (30%, n = 11). A 2×2 contingency table was constructed using categories for high and low mean CFQ scores against UR and IR, for which a chi-square test showed no significant association ($\chi^2_{(1)} = 0.79$, p = n.s).

6.3.2.2 Diary Responses

Question 1 provided context to confirm no significant difference in operating environment existed compared to the initial study conducted in Chapter 5. Thus, intentionally, no qualitative analysis was attempted. Questions 2 and 7 were used to calculate the time (T) between the error (e) and latent detection (d), recorded as T(e-d), for which Table 6.2 provides a summary of mean times for T(e-d) against location. A Kolmogorov-Smirnov test on timing data showed $D(51) = 0.36$, p < 0.05, indicating the distribution of the times for T(e-d) deviated from a normal distribution (positive skew = 2.58). The distribution gave a mean time for T(e-d) = 120 mins (n = 51, sd = 216, se = 30, range 2–1020 mins) with a mode of 30 mins and median of 30 mins. Notably, 78% (n = 40) I-LEDs occurred within 120 mins.

Question 3 recorded past errors associated with either complex or simple tasks. For example, a complex task included maintenance activities such as rigging flying controls and in-depth fault diagnosis. Examples for simple tasks included checking oil

TABLE 6.2
Error Locations

Location (n = 51)	Response	Count	Mean Time T(e-d) (min)	Example
Q6: Location of	AMCO	19		
error occurrence	Hangar	8		
	Line	3		
	Maintenance office	2		
	Issue centre	8		
	Storeroom	3		
	Aircraft	4		
	Workshop	0		
	Flight deck	2		
	Other	2		Crew room, head office
Q8: Location of	AMCO	15	106	
I-LED (ungrouped)	Hangar	9	24	
	Line	3	19	
	Maintenance office	1	15	
	Issue centre	3	45	
	Storeroom	0	–	
	Aircraft	3	28	
	Workshop	0	–	
	Flight deck	3	23	
	Crew room	2	45	
	Home/Mess	5	273	
	Bed	1	420	
	Vehicle	2	392	
	Gym	0	–	
	Other	4	324	Locker room, briefing room, bar (x2)

levels, returning tools, basic data entry tasks and logistics activities such as sending and receiving stores. Participants reported 14% (n = 7) complex tasks and 86% (n = 44) simple tasks. Responses to Question 4 indicated 92% (n = 47) were event-based, 2% (n = 1) task-based and 6% (n = 3) were recorded as both. For event-based activities, 64% (n = 30) were associated with UR, and 36% (n = 17) were associated with IR, for which 71% (n = 12) of this group reported against Question 11 that their IR was part of a personal routine (not part of a mandated procedure). A 2 × 2 contingency table was constructed using the categories for review type against the main cue for the task carried out in error. Due to low frequencies, a Fisher's exact test was carried out, which showed no significant association (P = 1.0, p = n.s). Counts for IR and UR are shown against each activity in Table 6.3.

Reason's (1990) GEMS taxonomy was used in Question 5, which expands on Norman's (1981) research that described error types based upon the incorrect use of schemata. Significantly more lapses were reported than mistakes and slips: 90%

TABLE 6.3
General Environment against Associated Factors

Environment (Q8)	Activity (Q9)	Review (Q10)		Work Thoughts (Q14a)			Non-Work Thoughts (Q14b)			Physical Trigger (Q13)	
		IR	UR	Past	In-Hand	Future	Past	Moment	Future	Item	Example
At work (n = 41)	Planning/preparing maintenance (n = 10)	6	4	4	4	2				Document = 5 Vista = 1 Other = 1	Aircraft documentation Scenery inside building None specified
	Conducting similar maintenance (n = 13)	5	8	2	9	2				Equipment = 4 Document = 6 Vista = 1 Sound = 2	Computer, rotor blades Aircraft documentation None specified Aircraft noise, headset volume
	Walking (n = 7)	2	5	1	2	3			1	Equipment = 5 Other = 1	Aircraft, tools, toolbox Felt keys in pocket
	Eating (n = 1)		1						1	–	None specified
	General work discussion (n = 3)	1	2	1		2				Other = 1	None specified
	Daydreaming (n = 3)		3	1	1			1		Vista = 1	None specified
	Resting (n = 1)		1					1		–	None specified
	Other (n = 3), that is changing clothes, paperwork and auditing		3	2				1		Equipment = 2 Document = 1	Screw bag Aircraft documentation

(Continued)

TABLE 6.3 (Continued)

General Environment against Associated Factors

Environment (Q8)	Activity (Q9)	Review (Q10)		Work Thoughts (Q14a)			Non-Work Thoughts (Q14b)			Physical Trigger (Q13)	
		IR	UR	Past	In-Hand	Future	Past	Moment	Future	Item	Example
Not at work (n = 10)	Driving vehicle (n = 1)		1						1	Sound = 1	Work-related topic on car radio
	Showering (n = 2)		2				1	1		Equipment = 1	Keys
										Other = 1	None specified
	Socialising (n = 2)	1	1	1				1		Vista = 2	None specified
	General work discussion (n = 1)		1	1						Sound = 1	Colleague's voice
	Resting (n = 2)	1	1	1				1		Document = 2	Aircraft documentation
	Sleeping (n = 1)		1	1						–	None specified
	Other (n = 1), that is readying for work		1					1		Other = 1	None specified
	Totals	**17**	**34**	**13**	**17**	**10**	**1**	**8**	**2**	**40**	

TABLE 6.4

Example Narratives against the GEMS Taxonomy (Reason 1990) and Norman's (1981) Schema-Related Error Types

Example Narrative	Erroneous Act (Error) Classification	
	GEMS (Reason 1990)	Schema Action (Norman 1981)
'I did not replace the oil filler cap correctly...had to go back and check'. 'Walking from the hangar to the flight deck I dropped a tool in my pocket'.	Slip	Correct intention selected but faulty schema(s) activation.
'Having prepared a Lynx [helicopter] for flight...the book [aircraft documentation] was completed. Just prior to launch, I realised that I hadn't cleared a Pt1 entry [statement of required maintenance]'. 'Forgot to fit main rotor spectacles [rotor blade securing device]'.	Lapse	Correct intention selected but schema(s) not triggered.
'Card raised [maintenance paperwork] for a maintenance task required post flying serial...incorrect aircraft annotated on paperwork'. 'Failed to co-ordinate [complete] a maintenance work order correctly'.	Mistake	Incorrect formation of intent. Wrong schema selected based on incorrect perception of external sensory data.

(n = 46), 6% (n = 3) and 4% (n = 2), respectively. Thus, the post-task detection of latent execution errors (slips and lapses) represented 94% of all maintenance tasks reported by participants, with planning errors (mistakes) accounting for 6%. Table 6.4 provides example narratives against the GEMS taxonomy and also Norman's (1981) schema-related error types for completeness.

Derived from Questions 6 and 8, Table 6.2 also provides a count for locations where the error occurred and I-LED event, and 73% (n = 37) error events were detected in the same or similar environment to that which the error occurred. Note that the AMCO is the Air Maintenance Coordination Office where most aircraft paperwork is controlled and maintenance organised. The Issue Centre is where ground equipment and tools are stored and controlled. The Line and Flight Deck are where aircraft operations are conducted, which is similar to a Ramp in civilian contexts. These locations could be grouped as environments At Work and Not at Work as shown in Table 6.3. This shows 80% (n = 41) I-LED events occurred whilst At Work and 20% (n = 10) whilst Not at Work. A 2 × 2 contingency table was constructed for environment against review type. Fisher's exact test showed no significant association (P = 1.0, p = n.s) although UR was dominant At Work (n = 27) and Not at Work (n = 7). Table 6.3 also shows activity at the time of recall, reported against Question 9. This indicates 25% (n = 13) of participants were engaged with similar maintenance task, 20% (n = 10) were planning or preparing to conduct maintenance task or 14% (n = 7) were simply walking (between activities). The remaining activities (n = 21) are highlighted in Table 6.3, noting that activities covering dissimilar maintenance, exercising and entertainment are not included, as participants reported none. Table 6.3 also shows 33% (n = 17) of participants intentionally reviewed past tasks.

Question 13 asked participants to record anything in the immediate physical environment that they believed might have triggered their error recall. Responses are shown in Table 6.3, for which aircraft documentation accounted for 35% (n = 14) and aviation equipment 30% (n = 12). A 2 × 4 contingency table was constructed for environment against triggers for Documentation (n = 14), Vista (n = 5), Sound (n = 4) and Equipment (n = 12). Other (n = 5) was not used, as participants did not provide examples. A Fisher's exact test showed no significant association (P = 0.42, p = n.s); however, 30% (n = 12) of all reported triggers were At Work and related to aircraft documentation followed by 28% (n = 11) aviation equipment.

Table 6.3 also highlights responses to Question 14, which asked participants to record what they were thinking at the time their I-LED event. 78% (n = 40) reported work-related thoughts and 22% (n = 11) non-work-related thoughts. Worked-related thoughts ranged from thinking about past maintenance activities to the task in hand through to maintenance to be carried out at a later time. Non-work-related thoughts ranged from past personal errands, simply existing within the 'moment' through to thinking about a personal task to be done later. Examples include a previous social event, in the moment watching TV or thinking about what Personal Computer (PC) game to play later. When an I-LED event occurred, 59% (n = 30) of participants were alone according to the responses to Question 15 (i.e., not actively engaged with another person such as talking or working on a task together), during which 87% (n = 26) occurred when carrying out largely autonomous tasks requiring little focussed attention. Aviation-related examples include planning and conducting simple maintenance tasks (n = 8) such as basic aircraft servicing, tool checks and simple logistic tasks. Non-aviation-related examples included walking (n = 7), driving a car (n = 1), showering (n = 2), daydreaming (n = 3), resting (n = 2), sleeping (n = 1) and other (n = 2) such as changing clothes and getting ready for work.

Question 12 indicated that all participants checked their work when a past error came to mind, for which 92% (n = 47) found the error to be real, whilst 8% (n = 4) experienced a false alarm. Figure 6.1 describes the distribution of responses (n = 50) to the Likert coding specified for Question 16, 17 and 18. Participants generally agreed 46% (n = 23) or strongly agreed 36% (n = 18) that their specific error was very clear to them. Participants were uncertain 26% (n = 13) or agreed 26% (n = 13) that they were very confident in their past task being in error, whilst 28% (n = 14) strongly agreed, and 30% (n = 15) strongly agreed that they were highly focused on the activity they reported at Question 9. This was closely followed by 28% (n = 14) who agreed, although 20% (n = 10) strongly disagreed.

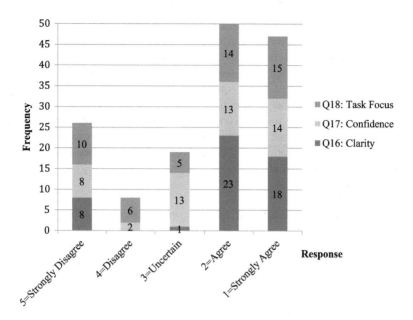

FIGURE 6.1 Responses to questions on task focus, confidence and clarity.

6.4 DISCUSSION

6.4.1 INDIVIDUAL FACTORS

6.4.1.1 CFQ Scores

CFQ scores were not used to predict I-LED events or propensity for human failings. The mean CFQ score for participants who returned their diary was similar to other studies of naval personnel (Wallace et al., 2002; Bridger et al., 2010). The low mean CFQ score indicated the sample of aircraft engineers possessed good executive function normally found in skilled workers (Broadbent et al., 1982). Studies have shown this helps cope with high workloads (Finomore et al., 2009; Bridger et al., 2010). However, participants described routine habitual tasks; thus, workload did not necessarily influence I-LED. Most I-LED events were associated with simple tasks. Arguably, the routine nature of everyday tasks infers low arousal. This may account for the high number of simple tasks, although good executive function should result in increased cognitive awareness. Thus, it can only be surmised that complex tasks tend to be more safety critical. Expectedly, an organisation's safety system attempts to defend critical tasks with more layers of error protection (Safety I: Hollnagel, 2014) than simple tasks. This perhaps results in fewer examples of I-LED involving complex tasks. A high CFQ score may indicate someone who is likely to be less receptive to external cues, against which behaviours may have been created that promote resilience. However, the sample possessed a low mean CFQ score, and since more URs than IRs were reported, results appear to support literature that reports cohorts of skilled workers are less likely to need to deliberately check their work (compared to less skilled workers with high mean CFQ score) (Reason, 1990; Mecacci & Righi, 2006; Day et al., 2012). However, the authors expected a greater number of IR than UR events as aircraft engineers operate in a safety critical environment. A possible explanation is that the reported errors were of a potentially minor nature and everyday experience (e.g., forgetting to return equipment, keys left in pocket, basic data entry errors etc.). Thus, general behaviours may exist within the cohort of naval aircraft engineers where everyday minor tasks are not considered to represent sufficient concern to warrant deliberate checking. Here, the systems approach looks to offer mitigation either by promoting cues to achieve I-LED or through organisational resilience to undetected errors.

6.4.1.2   Error Type

Lapses were associated with most I-LED events. Error research often reports the dominance of lapses, which offers an account for the overall high number of execution errors (Woods, 1984; Blavier et al., 2005; Reason, 2008; Kontogiannis & Malakis, 2009; Saward & Stanton, 2015b). The low count of I-LED events involving slips was expected as action errors tend to be detected proximal to the error event due to the transparency of the action (Reason, 2008), for which the action-world element of perceptual cycle also supports this expectation, as action errors should be readily apparent during performance monitoring. Arguably, mistakes tend to be less evident to attentional mechanisms due to the implicit lack of awareness of the genotype/phenotype schema mismatch for a given task, which may account for the low count in I-LED events associated with planning errors. Therefore, the post-task detection of mistakes is likely to be problematical for the PCM since important cues are simply not recognised (Norman, 1981; Plant & Stanton, 2013b). This incorrect perception of the world element of the PCM creates a lack of situational awareness. When situational awareness is absent during the perceptual cycle, it is argued that schema housekeeping is ineffective unless an important cue is recognised later. Thus, reviewing past tasks can lead to a regain of PCM situation awareness as the genotype/phenotype schema mismatch between intent and the external world is identified from system cues that trigger error recall. The fact that this activity occurs after a task is completed may add to PM theories such that the internal marker or 'to-do' list is not deleted from memory upon completion; otherwise, there would be no reference against which to conduct schema housekeeping. Also, the detection of past errors without a conscious attempt to review past tasks appears to confirm the presence of the SAS and an autonomous capability in schema housekeeping over time, although it is unclear whether this autonomy is systematic or simply dependent on cue recognition and is therefore indiscriminate.

6.4.2   SOCIOTECHNICAL FACTORS

6.4.2.1   Time

Most I-LED events occurred within two hours of the error event, which is a similar finding from the study in Chapter 5. This is an important finding as Plant and Stanton (2013a) highlighted PCM theory involves a schema-action-world cycle, for Chapter 5 argued that the perceptual cycle persistently reviews past task performance, and thus, time must be associated with the frequency of this cycle, which facilitates a 'golden window' in which most work-related I-LEDs occur. Although the standard deviation for the current study was large, indicating the distribution of mean times would benefit from a larger sample to improve statistical confidence, timing data represent real-world evidence. Additionally, the correlation with a separate sample of aircraft engineers offers confidence that there appears to be a relationship between time and the perceptual cycle, which can be persistent. T(e-d) for IR was found to be almost three times quicker than UR, indicating deliberately focused PCM attention can identify a latent error condition more quickly than the autonomous condition. In each case, I-LED events demonstrate information about a completed task remains accessible to the perceptual cycle, even though the intended task may have been removed from the internal 'to-do' list. Here, it is perhaps autonomous schema housekeeping that is responsible for continued engagement with the perceptual cycle over time, although deliberate intervention via an intentional review is also effective. If there is indeed a delayed cycle time associated with I-LED, then accommodating this delay in maintenance planning may reduce the likelihood of undetected errors remaining in the system during flight. However, this would be a challenging intervention in a real-world operating context.

6.4.2.2 Environment

I-LED is thought to be most successful when schema housekeeping takes place in the same or similar environment to that which the error occurred (Saward & Stanton, 2015b). In the current study, most I-LED events occurred At Work in the same or similar location to that which the error occurred (e.g., an error occurring in the hangar and was later recalled in the AMCO). This seems reasonable as most I-LEDs occurred within a time window of two hours whilst the engineers were still working. Work-related I-LED events also occurred when Not at Work, although there were fewer reported, but this does provide evidence that the PCM is able to remotely link to workplace cues to detect past errors, despite being physically present in a different sociotechnical environment. However, there are significantly greater delays to I-LED when Not at Work, as highlighted in Table 6.2. In most cases, though, error recall was associated with a cue. This demonstrates the importance of system cues distributed across environments to trigger recall.

6.4.2.3 Cues Triggering Recall

Participants reported familiar and recognisable workplace cues that they perceived to trigger schema recall. At Work, participants reported physical work-related cues such as sounds, aircraft paperwork, ground support equipment, toolbox and so on. Not at Work, participants reported physical items such as keys or a screw bag but also indirect perceptions of work-related cues. For example, a participant reported thinking about aircraft paperwork at home before realising an entry was made in error, another participant reported holding a general discussion about work with a colleague when a past error came to mind, and one via an aircraft-related news item on the radio. This appears to highlight the perceptual cycle's dependency on cues to trigger I-LED, which can involve the internal visualisation or phonological review of work-related cues in addition to physical cues away from the workplace. If the perceptual cycle can access representations of work-related cues when not in the workplace, either by visualising or reconstructing past activity internally, it is therefore argued important cues must remain available through residual memories and be sufficiently meaningful to trigger any genotype/phenotype schema mismatch. This leads to successful I-LED (Reason, 1990; Stanton et al., 2009b; Saward & Stanton, 2015b). Kvavilashvili and Mandler (2004) found memories of past events were mostly triggered by easily recognised physical cues in the external environment, which are mostly visual or auditory. In the current study, At Work aircraft documentation and aviation equipment were prevalent, whilst only two occasions of sound-related triggers were reported whilst At Work (aircraft noise and headset volume). Further, Mazzoni et al. (2014) argued that written word cues are more likely to trigger past memories (and thus I-LED) than picture cues. Although theirs was a laboratory-controlled experiment, current data appear to support this position since written words are implicit with aircraft documentation. Of note, the picture cues they employed were card cues and not physical objects encountered in the workplace. The only work-related pictures that the authors found in the workplace (apart from technical diagrams in maintenance manuals) are those on flight safety posters, but participants did not report I-LED events involving posters; thus, no further analysis could be attempted.

For I-LED events where no physical trigger was perceived, this could indicate that the participant was simply not aware of the trigger or confirms autonomous schema housekeeping, perhaps giving rise to chance detections. In this situation, the general environment may be the trigger as opposed to specific cues. For example, Sellen (1994) found that intentions are often recalled due to contextual factors rather than spontaneous retrievals. Thus, simply being immersed in an associated environment or recreating associations internally may aid I-LED.

Few I-LED events were associated with a false alarm, which suggests schema housekeeping often identifies latent error conditions correctly. Here, participants mostly agreed or strongly agreed that their error was clear to them, although they were slightly less confident that the past task was actually in error. For example, 'I don't think I replaced the oil filler cap correctly, but did I?' Arguably, this may imply weaknesses can lie in the transactional relationship between schemata and the action-world element of the PCM rather than the actual schema being faulty. I-LED events often occurred working alone on another task, which requires little focussed attention. This seems to suggest the PCM has capacity to attend to schema housekeeping, even when attending to an ongoing task, which may limit false alarms provided other people do not distract the engineer. Few participants thought about work-related topics when Not at Work, whilst most reported thoughts 'in the moment' and 'task in-hand'. Both are argued to be largely unfocussed activities associated with autonomous behaviour and therefore may offer further evidence that cognitive capacity is afforded to schema housekeeping during these periods of activity. This should be expected according to Kvavilashvili and Mandler (2004) who found memory recall is most likely during periods of unfocussed attention either during periods of inactivity or engaged in largely autonomous activities that do not require high levels of concentration on the ongoing task. This risks a lack of cue recognition needed to assure situation awareness of the perceptual cycle proximal to the error event but also seems to suggest the PCM has in-built capacity to attend to an ongoing activity whilst conducting schema housekeeping. This argument appears to receive support from Wilkinson et al. (2011) who similarly found expert operators exhibit the capacity to respond to cues related to the ongoing task whilst also engaging with the external environment for error checking. That said, Figure 6.1 shows participants also reported they were highly task focussed when the I-LED occurred. Since all reported tasks are argued to be routine (see Table 6.3), it is considered more likely that they were not highly focussed in terms of cognitive demands but more that the participant simply means it was the only task they were engaged with.

6.5   I-LED INTERVENTIONS

The significant number of I-LED events associated with lapses may indicate weaknesses in a cues ability to trigger the intended schema action (Norman, 1981). Thus, safety-focussed organisations should consider the 'strength' of existing cues as an intervention to help enhance an error-detecting environment through assured engagement with the perceptual cycle. If physical objects (aviation equipment) and word cues (aircraft paperwork) mostly influence accurate I-LED, this may offer other avenues of engagement with the perceptual cycle, for example, specific word cues strategically placed alongside data entry areas in aircraft paperwork (i.e., the maintenance process or signature sheet) such as 'check units', 'panels', 'keys', 'tools' and so on. This deliberately targets the triggers reported by participants. Aircraft paperwork (hard copy or electronic paperwork) is often completed in a separate location to where actual maintenance is conducted; therefore, placing related paperwork near to where maintenance or supporting activities are carried out may improve transactional relationships, potentially reducing error initiation and/or enhancing I-LED, for example, signing for aircraft work as near to the aircraft as possible by moving necessary paperwork to the aircraft or issuing the air engineer with a portable e-tablet so specific task elements can be intentionally reviewed whilst remaining immersed in the same physical environment, thereby avoiding dissociation of sociotechnical context. Further, messages on flight safety posters could be replaced with simple images of relevant objects or perhaps use small display stands positioned in locations such as the AMCO or Line. Displaying actual physical objects associated with common errors could enhance cue recognition, for example, an oil dipstick, padlock and key, fuel filler cap or indeed a scaled model of the aircraft. A formal 'stop and check' of simple tasks, even for minor tasks, may offer effective intervention. This is thought to be most effective if conducted within two hours of a completed task and alone to avoid distraction, and if conducted near where the task was executed. Interventions are considered especially important for simple everyday habitual tasks carried out alone. Here the 'Stop, Look and Listen' strategy used in UK road safety campaigns could be applied to air safety. 'Stop' refers to the PCM cycle time, 'Look' refers to sensing physical cues or the internal visualisation of past tasks, and 'Listen' refers to phonological cues that could simply include the internal voicing of the 'to-do' list. Importantly, the golden two hours window in which most I-LED events occurred is an intervention on its own, but clearly, any of the interventions previously described are likely to shorten detection times (depending on when the intervention is initiated).

6.6 STUDY LIMITATIONS

Naturalistic research is clearly challenging and limits the data that could be collected, but ecological data was essential to gain the necessary insight into erroneous acts with successful recoveries in the workplace, for which it is believed the current study has advanced understanding on I-LED. Analysis was limited to 51 usable I-LED events, which may mean I-LED is not as prevalent as thought or that the two months given to complete the diary was not long enough for more I-LED events to occur. Feedback from squadron engineering management is thought to provide a further explanation for the low return. It was highlighted that participant workload was very high, so they were not always able to make diary entries whilst several engineers were re-employed away from the squadron or on a short-notice course. Additionally, the self-report diary approach may have biased the count of I-LED events due to increased vigilance. Thus, future I-LED research would benefit from study of a larger sample over a longer period in a cohort that is able to commit fully to completing diary entries. Since the study was limited to a population of highly skilled engineers, a sample of unskilled workers should be considered as well as scenarios involving less familiar tasks and/ or high workloads. In the study of human performance variability, only the context changes (Robson, 2011; Cheng & Hwang, 2015); thus, it would be advantageous to conduct I-LED research in other workplace contexts where there are clearly more sociotechnical factors of interest than could be covered in the current study.

It was beyond the scope of the current study to report the safety risk of not detecting the latent error conditions highlighted in the current study. To make this assessment requires hierarchical task analysis and accident causation modelling to explore how particular undetected errors or erroneous acts might contribute to a safety occurrence. However, the potential benefits of integrating I-LED interventions as additional safety control within an organisation's existing safety system are discussed in Chapter 9. Modelling of the entire At Work and Not at Work environments to report frequencies of all complex/simple tasks carried out by the engineers in all locations where erroneous acts occurred and later recalled was also beyond the scope of the current study. Participant high workloads and the safety critical nature of the observed squadrons did not permit this additional data collection, as more extensive diary questions and/or separate observations would have been necessary, that is participants would need to report all complex/simple tasks carried out each day, in addition to 24/7 tracking of participant movements to report location frequencies. Thus, the risk/benefit of I-LED interventions modelled against frequencies for all possible maintenance-related activities would warrant separate research.

6.7 SUMMARY

The aim of this study has been to advance knowledge of the nature and extent of I-LED events from a system perspective and to progress Objective 3, which is identify practicable interventions that enhance I-LED events in safety critical contexts. A diary study was used to observe naval aircraft engineers in the natural workplace, which was designed around a multi-process approach to systems research was used that combines theories on PM, SAS and the schema theory within the PCM.

Additionally, the CFQ was administered to simply affirm that the sample exhibited normal cognitive behaviours associated with skilled workers; thus, the current findings are likely to be transferrable to other populations of skilled workers. Previously unreported I-LED events appear to show successful safety behaviour (Safety II events) to be effective upon the deliberate review of past tasks within a golden time window of two hours of the erroneous act occurring. Notably, this occurs during periods of unfocussed attention and whilst working alone in the same or similar sociotechnical environment to that which the error occurred. Several sociotechnical factors associated with I-LED were studied so that practicable interventions could be identified, which are anticipated to enhance I-LED and therefore contribute to safety barriers in the workplace. Application of these practicable I-LED interventions using a systems approach is considered especially important for simple everyday habitual tasks carried out alone where perhaps individual performance variability or human error effects are most likely to pass undetected if there are deficiencies in an organisation's safety controls or defences. It has been argued that I-LED interventions are likely to offer further resilience against human performance variability by helping to regain SA within the perceptual cycle through deliberate engagement with system cues; particularly physical objects such as equipment or written words. However, it is recognised that the interventions identified in this study need to be deployed within naturalistic real-world contexts to operationalize and test their true benefit in terms of risk mitigation, frequency and effectiveness over time. By definition, I-LED compliments Safety II events by supporting individual safety behaviour, yet any I-LED intervention is also likely to support Safety I control strategies and thus should be integrated within an organisation's safety system. Thus, it is believed safety-dependent organisations should look to improve their safety system using I-LED intervention techniques that deliberately engage with system cues across the entire sociotechnical environment and full range of normal workplace behaviours to trigger recall, as opposed to chance detections, thereby providing opportunities for enhancing safety barriers as mitigation for system-induced human error effects.

The potential I-LED interventions identified satisfy Objective 3 in the current safety research and will be tested in Chapter 7 using a new cohort of aircraft engineers who are observed in their everyday working environment. The study aims to understand which I-LED interventions deliver the greatest safety benefit using system cues available in the working environment and is therefore designed to address Objective 4, which seeks to understand the effectiveness of I-LED interventions in the workplace.

7 I-LED Interventions
Pictures, Words and a Stop, Look, Listen

7.1 INTRODUCTION

Chapter 2 considered some of the many definitions covering the term *human error*, from which it was argued human error is simply a colloquial expression used to flag error effects or erroneous actions that must be contextualised and explained against system causes (i.e., situational error) to be meaningful. Ambiguity in definitions, leading to a misunderstanding of error analysis, was explored further in Chapter 3 to address concerns that the term is outdated when analysing safety failures within complex sociotechnical systems. New terms such as erroneous acts, human performance variability or system failures have emerged to describe error effects associated with human activity where the real causes of safety failures are deep-rooted in system factors such as organisational decisions, design, equipment, management oversight and procedures (Woods et al., 2010; Dekker, 2014; Stanton & Harvey, 2017). Application of a systems perspective opens a more productive dialogue on performance variability that includes normative and non-normative behaviours and therefore a need to design resilient workplace safety systems. This encompasses an operator's ability to self-monitor for system hazards (traps) and correct as necessary to help manage safety at a local level in the workplace (Stanton & Baber, 1996; Reason & Hobbs, 2003; Woods et al., 2010; Reiman, 2011; Cornelissen et al., 2013). Chapter 3 argued the term human error can survive as a valid descriptor in systems safety but only if it is used carefully to highlight the need to analyse the causal effects of safety failures generated by the system and not by the individual. For the purpose of the current safety research, human error that passes undetected becomes a latent error condition, which can impact future safety performance (Reason, 1990). Here, the term refers to the residual effects created when the required performance was not enacted as expected due to system-induced sociotechnical traps generated by the organisation, that is system failures that pass undetected and therefore lie hidden (Reason, 1990).

Examples of everyday failures might be leaving the gas hob on when leaving home or failing to lock the door of their house. Both could have potentially negative consequences, if left undetected. Arguably, most people have experienced the phenomenon later and, spontaneously, recalling that the gas hob has been left on or they failed to lock the front door. This book focusses on aircraft maintenance where common examples include the later realisation that a tool was not removed from the aircraft engine bay, an oil filler cap was not replaced after replenishing the reservoir, or the aircraft documentation wasn't completed correctly. These typical examples of maintenance provide the catalyst to design practicable system interventions for use in the aircraft maintenance where the timely and effective detection of latent error conditions can make maintenance safer and, in turn, more effective. The following chapter explores the effectiveness of a several Individual Latent Error Detection (I-LED) interventions applied to the workplace. This addresses Objective 4 in the current research, which is to understand the effectiveness of I-LED interventions during normal operations in the workplace.

Chapter 5 highlighted that the I-LED phenomenon has been observed where errors suffered by naval aircraft engineers at work appear to be later detected spontaneously by the individual at some point post-task completion, without reference to recognised procedures. Chapter 6 found I-LED to be most effective when engaging with system cues that trigger recall within a time window of two hours. Detection appeared to be improved whilst the engineer worked alone in the same environment that the error occurred, particularly if physical cues such as equipment and written words were present. This suggests a level of safety exists within the workplace that has not previously been accounted for in organisational safety strategies. Human error is often quoted as contributing to 70+% of accidents (Helmreich, 2000; Wiegmann & Shappell, 2003; Adams, 2006; Flin et al., 2008; Reason, 2008; Woods et al., 2010; Saward & Stanton, 2017), but this belies systemic causes that do not adequately control or manage human performance variability in achieving workplace safety (Leveson, 2004; Morel et al., 2008; Amalberti 2013). I-LED research adopts the systems perspective where it is system cues that trigger recall but from a human-centred approach (Stanton & Salmon, 2009) to reveal understanding of how individual acts of post-task error detection contribute to total safety within complex sociotechnical systems. This involves the interaction between humans and technical aspects of the environment such as equipment, technology and workplace processes (Walker et al., 2008; Niskanen et al., 2016).

The step-change from studying error as a causal attribution of blame to a symptom of wider systemic issues has led to a paradigm shift in the aetiological approach to safety performance or total safety using systems thinking (Leveson, 2004). Little is known about individual error detection (Blavier et al., 2005; Saward & Stanton, 2015a,b), although it is argued I-LED can offer a further shift in safety thinking. The phenomenon addresses everyday errors that could be considered insignificant but where accident causation modelling later revels complex paths of converging latent error conditions within the system as a whole. It is argued safety is created through managing risks by controlling hazards (system traps) that can cause harm. The management of risks posed by system hazards is represented in the I-LED model introduced in Chapter 4 and encompasses all system-induced operator errors, regardless of perceived significance. Morel et al. (2008) observed total safety is the product of controlling safety risks (system controls such as rules and procedures, training and experience, supervisory controls, etc.) and managing safety risks locally (through the adaptive capabilities of operators within system controls). Therefore, it is believed that the safety aim of an organisation should not be preventing all errors occurring but more towards using a systems approach to risk management of latent error conditions by promoting resilience using a total safety approach This approach encompasses Situational Awareness (SA) regain through I-LED events discussed in Chapter 5, especially where safety control mechanisms are exhausted through exceptional conditions (Hollnagel et al., 2006; Amalberti, 2013; Chatzimichailidou et al., 2015; Saward & Stanton, 2017). This can include occasions where operators find rules and procedures are ineffective or unavailable for a task, equipment is poorly designed or not available or organisation-driven error-promoting conditions such as fatigue, task pressure, workplace distractions and so on.

Kontogiannis (2011) demonstrated that error detection could be used in the design of error tolerant systems, which supports resilience and contributes to the mitigation of system-induced error effects to help assure total safety in the workplace. This view is similar to Hollnagel's (2014) modelling of accident causation, which highlighted Safety II events where safety is managed effectively at the local level in complex sociotechnical environments despite a myriad of system influences on human performance. Here, it is essential that the operator possesses error-detection skills in a working environment that promotes the cues needed to detect and recover from system-induced errors (Cornelissen et al., 2013). I-LED is a Safety II strategy aimed at supporting operator detection of their latent error conditions post-task completion. Thus, the current safety research does not consider error prevention but the management of operator engagement with system cues to help support the timely detection of past errors before they propagate and combine with other factors to become an accident (Reason, 1990). For example, Amalberti (2013) noted that routine error rates can be high, but the true safety performance of a safety critical organisation should be judged against the rate of detection and recovery, since the risk of error comes from its consequences if not intervened early. He noted that, in addition to established safety rules and procedures, the safest hospitals are those with the overriding ability of its operators to detect their errors before an unwanted consequence occurs. It is argued that a safer aircraft maintenance environment is similarly one in which its operators possess effective I-LED skills.

The I-LED study described in Chapter 6 found system cues such as time, location and other sociotechnical factors that are present within the workplace and other environments such as at home could trigger successful I-LED. Their findings were based on a research using schema theory, which describes information represented in memory about our knowledge of the world we interact with to carry out actions (Bartlett, 1932). The associated schema-action-world cycle is characterised by the Perceptual Cycle Model (PCM), which describes the transactional relationship between the operator and system cues in the external world (sociotechnical environment) that trigger intended actions (Neisser, 1976; Norman, 1981; Mandler, 1985; Stanton et al., 2009a; Plant & Stanton, 2013a). The execution of an action requires the bottom-up processing of information from system cues in the world against top-down prior knowledge from memory (schema) to enact the action successfully (Neisser, 1976; Cohen et al., 1986; Plant & Stanton, 2013a). It is important to note this function since I-LED relies upon system cues to trigger a review of past schema-action-world cycles to determine the success of previous actions (Saward & Stanton, 2017). Visual cues are particularly effective cues to trigger I-LED events (as opposed to other senses) where written word cues and physical objects have generally been found to be more likely to trigger recall than picture cues (Kvavilashvili & Mandler, 2004; Mazzoni et al., 2014; Saward & Stanton, 2017). Chapter 6 argued human factors and ergonomics

(HFE) designed I-LED interventions that make use of physical objects and written word cues as well as a 'Stop, Look and Listen' (SLL) approach are most likely to be effective for I-LED. For the SLL approach, 'Stop' refers to pausing ongoing activity to facilitate a review by the PCM, 'Look' refers to sensing physical cues, written words or the internal visualisation of past tasks, and 'Listen' refers to phonological cues from internally 'voicing' activity associated with past tasks or simply listening to sounds in the external environment.

Amalberti and Wioland (1997) showed errors suffered by skilled operators can be frequent, whilst experience improved an operator's ability to detect more of their own errors due to an enhanced 'capacity' to detect important cues present in the external environment (Blavier et al., 2005; Wilkinson et al., 2011). This study observes a new cohort of naval air engineers in the workplace that are grouped by experience: junior 'operatives' and more experienced 'supervisors'. It is thought the supervisors in this study would commit more errors than the operatives yet detect more of their own errors. Further, any I-LED intervention would improve the self-detection of past errors due to the deliberate schema-action-world review of past actions. Word cues were thought more likely to trigger recall than pictures for supervisors (Kvavilashvili & Mandler, 2004; Mazzoni et al., 2014) as they spend more of their time managing maintenance documentation than operatives. Finally, the SLL intervention was hypothesised to be the most effective I-LED intervention for both operatives and supervisors since the technique is arguably the only intervention to be observed that promotes the review of past actions using internal cues in memory and physical objects in the sociotechnical environment. This offers the potential to maximise the PCM's I-LED capability. Chapter 2 argued that the PCM also exhibits an autonomous schema 'housekeeping' function where the routine monitoring of the schema-action-world cycle already provides a level of error checking and is also used to collect feedback from completed actions to facilitate learning and the acquiring of experience. This housekeeping function is thought to explain why I-LED events were reported by previous cohorts of naval aircraft engineers where past errors were recalled within a time window of two hours of the error occurring, and thus, it was anticipated the control groups described in the method would also experience I-LED events without an intervention applied.

7.2   METHOD

7.2.1   PARTICIPANTS

The Royal Navy Air Engineering and Survival Equipment School (RNAESS) was selected for the current study as it provided an accessible, safe and controlled environment in which to observe I-LED events. Here, two training squadrons exist, which emulate operating squadrons using aircraft and standard maintenance procedures and is therefore representative of the real-world environment. One squadron provides maintenance courses to operatives, and the other provides maintenance courses to supervisors. An operative is junior in rank and authorised to conduct simple aircraft maintenance tasks such as aircraft flight servicing and other supervised tasks. A supervisor is more senior in rank and authorised to carry out more complex maintenance tasks such as in-depth aircraft fault diagnosis, coordinating aircraft documentation and leading maintenance teams. Participants comprised 120 naval air engineers attending maintenance courses during the period May 2016 to February 2017. The sample included males (n = 108) and females (n = 12) in two groups of 60 (supervisors and operators). The low count of females is consistent with the population. Combined (supervisors and operators) mean age = 24.92 (sd = 4.1, se = 0.37, range = 17–38). Supervisor group mean age = 27.43 (sd = 3.02, se = 0.41, range = 23–38) and the operator group mean age = 22.42 (sd = 3.47, se = 0.45, range = 17–30).

7.2.2 Design

The study was piloted using representative courses of 12 operatives (comprising nine males and three females) and 12 supervisors (comprising 11 males and one female), whom did not form part of the main study. No significant issues were highlighted, and the pilot confirmed the RNAESS was a suitable environment in which to observe I-LED events.

Instructor availability to conduct observations and the additional constraint that each intervention had to be simple and quick to complete, to avoid impacting ongoing training, resulted in a maximum of four interventions and a control condition that could be tested in the current study. Based on the findings from the study conducted in Chapter 6, the four interventions were designed using picture cues, word cues and SLL approach with the dependent or outcome variable for this study being an I-LED event.

RNAESS instructors were consulted and a review of literature carried out to identify system cues most associated with typical maintenance errors (Latorella & Prabhu, 2000; Wiegmann & Shappell, 2001; Hobbs & Williamson, 2003; Reason & Hobbs, 2003; Liang et al., 2010; Rashid et al., 2010; Saward and Stanton, 2015a). Twenty cues were identified as a manageable number to include in a simple booklet of flashcards. Where necessary, cues were contextualised for the naval aircraft maintenance environment and tailored to reflect differences between operative and supervisor roles, as shown in Table 7.1. Each cue was represented as a word or picture, as well as a combination of both the word and picture for a particular cue. Examples are included in the Appendices: Appendix A shows a flashcard picture of the torch highlighted in Table 7.1; Appendix J shows the MF700C; Appendix K shows a combination of picture and words for the oil filler cap; and Appendix L shows a combination of picture and words for maintenance checks cue. The word, picture or combined cues were compiled separately as flashcards in A5 booklets. Each booklet comprised 20 flashcards containing one of the 20 word cues on each page in bold black print (Arial text, 72 point) or picture (non-complex colour image on a plain background, 6" × 4") or combination of the word and cue. Eight separately numbered booklets were produced for each of the three intervention techniques (SLL did not need a booklet). Each booklet contained exactly the same words or pictures, but the order cues appeared was randomised within each booklet to remove ordering effects and to help reduce participants learning the sequence of words and therefore becoming desensitised.

Representative practical tasks were selected through further consultation with the RNAESS instructors, which encompassed the following general aircraft maintenance

TABLE 7.1

System Cues and Practical Tasks for Operatives and Supervisors

System Cues (Explanation Given in Brackets Where Applicable)		Practical Tasks (Typical, Well-Practised Maintenance Tasks)	
Operative	**Supervisor**	**Operative**	**Supervisor**
Toolbox	Toolbox	1. Aircraft flight servicing	1. Specifying independent checks
Padlock	MAP (Military publication)	2. Air system charge	2. Component removal
Keys	Keys	3. Oil replenishment	3. Hangar brief and checks
Screw bag	PPE (Personal protection equipment such as goggles, mask, gloves, etc)	4. Component Torqueing	4. Coordinate aircraft paperwork in GOLDEsp
Dipstick	Questions (part of supervisor process)	5. Aircraft jacking	5. Component receipt and despatch checks
Panel	Panel		
Filler cap	Maintenance checks		
Circuit breaker	Circuit breaker		
MF731 label (paper tag showing component serviceability)	MF731 label		
Torque wrench	Torque wrench		
Aircraft 700C (aircraft maintenance paperwork)	Aircraft 700C		
Tool tally	FOD (Foreign object debris)		
Socket	Socket		
Cowling	Cowling		
Tyre	Hangar checks (safety procedures)		
Aircraft	Aircraft		
Pen	Pen		
GOLDEsp (e-database for paperwork)	GOLDEsp		
Torch	Torch		
Lubricant	Lubricant		

categories: aircraft documentation/paperwork, logistics tasks, aircraft servicing and aviation support tasks. Specific tasks were identified to reduce variance, and they were also tasks that the participants would carry out regularly during their course. This allowed the interventions to be tested on well-practised tasks that were observed at the end of each course during consolidation periods, which arguably limited any significant effects due to early stages of learning (Fitts & Posner, 1967). For each practical task shown in Table 7.1 (five for operatives and five different tasks for supervisors to allow for differences in their employment), the instructors carried out basic error analysis with the authors to identify the potential number of erroneous acts for each task.

7.2.3 OBSERVATIONS

Within the period of the study, five operative courses and five supervisor courses were available, during which five practical tasks per participant per course only could be observed due to available resources and to avoid disturbing ongoing training. RNAESS instructors were trained by the authors to conduct the observations, as it was not possible for the authors to be present every day over the period of the study. Each group was allocated 12 participants to one of the five intervention categories. Each course was loaded with more than 12 engineers; thus, there was sufficient redundancy to ensure the required observations could be achieved. The first course for each group acted as a control where instructors observed participants without a deliberate intervention applied. The four interventions were then introduced separately in the subsequent courses. This approach simplified data collection and helped removed biasing due to potential cross contamination between interventions. For each intervention, including the control, participants were observed over five tasks; thus, a total of 600 observations were recorded.

7.2.4 PROCEDURE

Participants received a brief on the study at the start of their course, and each participant was provided with the Participant Information Sheet (PIS) shown in Appendix H. Each participant was issued a participant number and completed a register that recorded their course number, gender and age. The instructors observed the tasks during consolidation periods at the end of their course. This procedure was adopted to help ensure the participant had sufficiently practised the task to become a learnt skill. The instructor discretely observed the participant carrying out the task then issued the intervention technique to the participant post-task completion. This was timed so that the technique was not issued immediately after completing the task but within the two hour I-LED window. The instructor selected a booklet at random and gave it to the participant, who was asked to work through all 20 words and/or picture flash cards whilst alone. If asked to try the SLL intervention, the participant was given a brief on the intervention before trying the technique. After the intervention, the participant returned to the instructor, who recorded their feedback using the observer form shown in Appendix M.

7.2.5 DATA ANALYSIS

Detection sensitivity theory can be used to highlight differences in hit rates and false alarms (Stanton & Young, 1999; Fawcett, 2006; Stanton et al., 2009b). For I-LED research, this is the difference between LED events leading to true latent error conditions being detected compared to false alarms or no recall at all, which allows the strength of effectiveness of each I-LED intervention to be determined. A 2×2

TABLE 7.2

2 × 2 Contingency Table for I-LED Signal Detection Calculations

		Latent Error	
		Yes	**No**
Recall	Yes	TP (Hit)	FP (False alarm)
	No	FN (Miss)	TN (Correct rejection)

contingency table can be constructed to determine the signal sensitivity or effect of each I-LED intervention, as shown in Table 7.2.

TP: Number of true positives observed (hit: where recall resulted in a true error detection).

TN: Number of true negatives observed (correct rejection: where no error was committed or recalled).

FP: Number of false positives (false alarm: where no error was committed but recall caused participant to check their work).

FN: Number of false negatives (miss: where an error was committed but not detected).

Matthews coefficient (phi) coefficient (Matthews, 1975) can be calculated from the binary values recorded in the contingency table using following equation for phi (Φ):

$$\Phi = \frac{(TP \times TN) - (FP \times FN)}{\sqrt{(TP + FP)(TP + FN)(TN + FP)(TN + FN)}}$$

A coefficient of +1 represents perfect positive correlation whereby the I-LED intervention led to all past errors being detected. A coefficient of 0 represents no correlation, whereas a coefficient of −1 indicates a perfect negative correlation.

7.3   RESULTS

A sift of returned data revealed that most of the observer forms shown in Appendix M were not completed fully, which meant that only data for Questions 1, 2, 6, 7 and 8 could be analysed.

Table 7.3 shows operatives experienced 144 errors, detected 45.8% (n = 66) and missed 54.2% (n = 78), whilst supervisors experienced 270 errors, detected 23.9% (n = 65) and missed 75.6% (n = 205). These findings are represented in Figures 7.1 and 7.2 across each intervention for operatives and supervisors. The detection sensitivity phi (Φ) for each intervention is also recorded in Table 7.3 and represented in Figure 7.3.

Figure 7.1 shows the operatives in the control group experienced some I-LED events, achieving 33% (n = 11) hits out of the total observed errors for this group without an intervention applied. Table 7.3 records a negligible phi value ($\Phi = 0.04$) for the control

TABLE 7.3

Observations for Operatives and Supervisors

Group	Sensitivity Factors	Intervention					Totals
		Control	SLL	Words	Pictures	Combined	
Operatives	Observed errors	33	22	22	39	28	144
	Hits (TP)	11	16	10	17	12	66
	False alarms (FP)	10	8	4	4	2	28
	Miss (FN)	22	6	12	22	16	78
	Correct rejection (TN)	24	34	37	27	34	156
	Phi (Φ)	0.04	0.55	0.41	0.33	0.45	0.36
Supervisors	Observed errors	53	29	103	33	52	270
	Hits (TP)	3	21	34	1	6	65
	False alarms (FP)	0	0	8	1	1	10
	Miss (FN)	50	8	69	32	46	205
	Correct rejection (TN)	36	39	11	38	29	153
	Phi (Φ)	0.15	0.78	−0.07	0.01	0.14	0.25

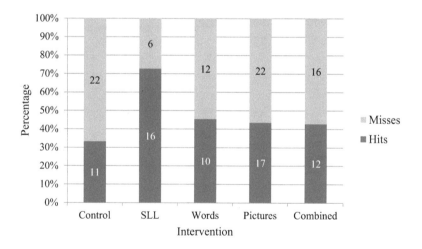

FIGURE 7.1 Percentage misses and hits for operatives.

group as these engineers experienced a similar number of false alarms (n = 10). The operatives who tried the SLL intervention achieved the most significant I-LED performance of all the operatives observed in the study with 73% (n = 16) hits, which aligns with the strong phi ($\Phi = 0.55$) for this group. Interventions using words, pictures and the combination of pictures and words all achieved improved I-LED performance compared to the control group, achieving 45% (n = 10) hits for words, 44% (n = 17) for pictures and 43% (n = 12) for combined (associated phi values shown in Table 7.3). Figure 7.2 shows the supervisors also experienced some I-LED events in the control group, achieving 6% (n = 6) hits out of the total observed errors for this group without

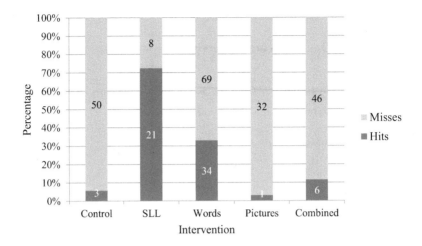

FIGURE 7.2   Percentage misses and hits for supervisors.

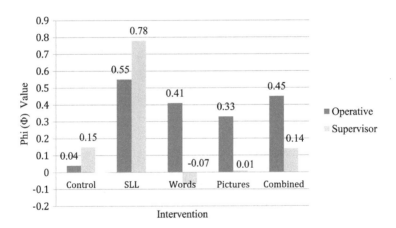

FIGURE 7.3   Phi (Φ) for each intervention.

an intervention applied. This result aligns with the very weak phi ($\Phi = 0.15$) recorded in Table 7.3, which is higher than the value recorded for operatives since the supervisors in the control group experienced no false alarms. The SLL intervention also achieved the most significant I-LED performance of all the supervisors observed in the study with 70% (n = 21) hits, which aligns with the very strong phi ($\Phi = 0.78$) recorded in Table 7.3 for this group. The value for phi is higher than for operatives, who achieved more hits (73%), as the supervisors did not experience any false alarms when using the SLL intervention. Supervisors achieved improved I-LED performance when using words and combined interventions, achieving 33% (n = 34) hits for words and 12% (n = 6) for combined. The phi value for the combined intervention was very weak ($\Phi = 0.14$), whilst the words intervention produced a negligible negative correlation ($\Phi = -0.07$), as shown in Table 7.3. Supervisors achieved 3% (n = 1) hits for pictures.

This is less than the 6% hits experienced by the control group, which also recorded a higher value for phi ($\Phi = 0.15$) than the supervisors using the picture technique ($\Phi = 0.01$), as shown in Table 7.3 and Figure 7.3. The results shown Figure 7.3 indicate the operatives experienced improved detection sensitivity compared to supervisors using the I-LED interventions for words, pictures and combined. These interventions had little effect on I-LED performance for supervisors compared to the control group, whilst the results for the operatives show these engineers all experienced improved detection sensitivity compared to their control group. The results in Figure 7.3 also show the superiors experienced a greater detection sensitivity using the SLL intervention than for operatives, with a strong phi value ($\Phi = 0.55$) for operatives and a very strong value for supervisors ($\Phi = 0.78$).

7.4   DISCUSSION

7.4.1   I-LED Intervention Performance

Errors committed by skilled operators can be high, yet they also possess the ability to detect more of their own errors due to an enhanced ability to detect important cues present in the external environment (Amalberti & Wioland, 1997; Blavier et al., 2005; Wilkinson et al., 2011). This study has shown supervisors committed almost twice as many errors as operatives, which only partially supports this position as the operatives detected a higher percentage of their past errors than supervisors (45.8% compared to 23.9%). Operatives and supervisors were equally expert for their tasks observed in this study, although the tasks for the supervisors were more complex than the simple maintenance tasks carried out by operatives. The study in Chapter 6 found that I-LED is particularly effective for simple habitual tasks; thus, this may explain why operatives experienced an overall higher detection sensitivity than supervisors when using an I-LED intervention ($\Phi = 0.36$ for operatives compared to $\Phi = 0.25$ for supervisors).

The results showed the presence of I-LED without an intervention applied, although the detection sensitivity was negligible for operatives and very weak for supervisors. This was expected as Chapter 6 reported that past errors could be detected within a two-hour time window if simply remaining immersed in the same environment to that which the error occurred. With an intervention applied, it was anticipated I-LED performance would see a significant improvement compared to the control due to the deliberate and targeted engagement with relevant system cues. Detection sensitivities for operatives showed a generally good improvement in I-LED performance across all interventions compared to their control group. The supervisors experienced negligible improvements in I-LED performance when using the words, pictures and combined interventions. Significantly, supervisor I-LED performance was worse than their control group, except for the SLL intervention.

The SLL intervention was effective for both groups of naval air engineers, which was expected as the intervention does not require significant focussed attention, compared to working through flashcards, and it is the only intervention that supports engagement with potential internal cues and external visual and auditory cues in the surrounding environment to trigger recall. The SLL intervention was particularly effective for supervisors, which may support the earlier argument that skilled operators possess an enhanced ability to detect important cues, but perhaps only if 'taking a moment' to reflect on their surroundings and thoughts rather than focussing on a document such as flashcards. Arguably, the activity of reviewing the flashcards may have caused a distraction during the schema-action-world cycle due to this focussed attention (Kvavilashvili & Mandler, 2004; Brewer et al., 2010; Rasmussen & Berntsen, 2011). This finding receives some support from Amalberti (2013) who noted that operator performance could be assessed through an individual's ability to detect and recover from error, which requires an element of self-reflection or metacognition, which is thought to further facilitate the schema-action-world cycle review of past tasks.

Operatives were also found to be generally responsive to directed engagement with cues found in the booklets, where the combination of pictures and cues was slightly more effective than pictures or words alone. Table 7.3 shows significantly more detections (hits) for supervisors using the word intervention. This should be expected as the sample of supervisors suffered approximately 4.5 times more errors than the sample of operatives who tried the word intervention. However, although the findings were not statistically significance, operatives experienced a greater detection sensitivity using word cues than supervisors ($\Phi = 0.41$ for operatives compared to $\Phi = -0.07$ for supervisors). This is a surprising result, as the supervisor role requires more time spent working with aircraft maintenance documentation and processes than physically working on aircraft. The word cues were carefully chosen to be contextually relevant; thus, the fact this intervention produced a slight negative effect may suggest the continued immersion in a 'word-rich' environment desensitises the supervisor to word cues, rendering the intervention ineffective for supervisors. The picture flashcards led to a substantial number of hits for operatives, whilst pictures and combined flashcards produced a significant detection sensitivity result for operators compared to supervisors, who experienced similar numbers of errors. The word flashcards resulted in a good detection sensitivity for operatives, though, which supports the view that written word cues are more likely to trigger recall than picture cues (Kvavilashvili & Mandler, 2004; Mazzoni et al., 2014; Saward & Stanton, 2017). Finally, the detection sensitivity for individual tasks for supervisors and operatives was also calculated for each I-LED intervention, including the control samples. No significant results were found, which indicated a latent error condition associated with a particular task was no more likely to be detected than for any other task.

Overall, the SLL intervention was found to be the most effective intervention for the aircraft engineers, with supervisors experiencing the greatest benefit. The other three interventions were ineffective for supervisors, showing similar performance to the control group. Operatives experienced an enhanced I-LED performance across all interventions compared to the control group. The findings generally support the position that visual cues can be effective triggers, for which the SLL intervention is likely to be the most effective I-LED intervention tested in this study.

7.4.2 I-LED Contribution to Safety-Related Risk Management

I-LED is argued to enhance safety-related risk management systems within safety critical sociotechnical contexts, as the interventions help guide operators to engage with system cues to achieve timely detection of past errors. This gives rise to Safety II events, which can also benefit to Safety I controls as part of a total safety approach (Morel et al., 2008; Amalberti, 2013; Hollnagel, 2014). Arguably, a resilient safety system is created from effective risk management both at the organisational and operator levels (Naderpour et al., 2014; Chatzimichailidou et al., 2015; Niskanen et al., 2016). Thus, it is argued I-LED is a practicable safety strategy due to its contribution to Safety II events, which should be integrated within an organisation's risk management system to enhance overall safety. This requires systemic changes to safety-related training and maintenance processes (Safety I) and the routine local application of I-LED interventions during normal operations (Safety II) to help mitigate for everyday workplace error effects such as those experienced by naval aircraft engineers, as described earlier. This is where the greatest benefit of the I-LED phenomenon is thought to exist, where routine habitual tasks can generate high error rates and where safety risks might be perceived as low by operators (Amalberti, 2013; Saward & Stanton, 2017). Arguably, I-LED can also help counter other potential consequences of undetected errors that are not safety-related, such as overall system performance, social-economic gains, and political and reputational value (Kleiner et al., 2015). Before introducing any I-LED intervention to enhance safety resilience, it is likely that the organisation will want to explore the cost versus benefit of integrating the additional control measure within their existing safety strategy, which is considered in Chapter 9.

7.5   STUDY LIMITATIONS

Observing and measuring multiple interacting sociotechnical factors during normal workplace operations to determine detection sensitivity with absolute confidence is challenging (Stanton & Harvey, 2017), and modelling of the entire network of sociotechnical environments (at work and not at work) to analyse I-LED events in all locations was beyond the scope of the study. This could limit the reported effectiveness of the interventions observed, although the two squadrons observed in the current study are closely related to other naval aircraft squadrons observed in Chapter 6 where data were collected to design the four interventions. In their study of prospective memory, Kvavilashvili and Mandler (2004) argued association priming improved memory recall. Thus, it could be argued that the association with simply using an intervention was the primer or trigger for general Latent Error Searching (LES) described in Chapter 2. The presence of general LES could account for the latent error detections seen in this study as opposed to specific cues contained in a booklet that triggered the recall of specific past errors. For example, did a picture of a pen trigger the recollection that a written record had been made incorrectly, or did simply taking time to read through the booklet provide a sufficient pause between maintenance tasks for routine schema housekeeping to occur? The current study attempted to collect data to account for general LES effects, but the absence of data recorded by the observers limited this ability.

7.6   SUMMARY

The aim of this study was to understand the effectiveness of I-LED interventions in the workplace, which satisfies Objective 4. I-LED has been shown to offer further mitigation for erroneous acts or system failures occurring in safety critical organisations, such as the aircraft maintenance environment observed in this study. Arguably, I-LED interventions are effective at enhancing the timely detection of past errors and, therefore, help to avoid adverse consequences such as a latent error condition networking with other latent factors to create a causal path to an accident. An effective intervention is context sensitive and maximises engagement with system cues. It is for this reason that the SLL intervention is likely to have been the most effective technique for both operatives and supervisors, as it immersed the engineers in their relevant sociotechnical environment. The SLL intervention is also a flexible technique that the operator can tailor independently during normal operations in the workplace. I-LED interventions should improve overall safety performance within an organisation, for which there are likely to be many more potential interventions than the four described. This may be especially true for habitual tasks carried out alone or for tasks perceived to be low risk where human performance variability could pass unchecked, with the potential for errors to pass undetected.

Organisational resilience comes from a safety-related risk management that matches human performance variability with system-based approaches. It has been argued safety is created through effective risk management that matches human performance variability with systemic approaches. The I-LED phenomenon is thought to offer a significant contribution to Safety II events, provided interventions are designed into Safety I strategy through training enhancements and safety processes. This requires a system perspective, as the organisation needs to embed I-LED interventions within its overall safety system to ensure operators receive the training, are given time to conduct the intervention, and context dependent cues are available in the workplace.

The I-LED phenomenon offers a further step-change in safety thinking by helping to manage system-induced human-error effects by facilitating Safety II events through the application of I-LED interventions post-task completion. Effective I-LED limits occasions for adverse outcomes to occur and can, therefore, help promote safety successes in the workplace. I-LED interventions applied to normal operations in the workplace should be of benefit to any safety critical organisation seeking to further enhance their existing safety system. The next chapter explores the integration of I-LED interventions within existing safety systems, whilst Chapter 9 assesses the benefits of this safety strategy.

8 A Total Safety Management Approach to System Safety

8.1 INTRODUCTION

Resilience is a safety strategy that is dependent on the effective management of hazards using system controls applied to the workplace that must be managed across the entire network of potential hazards with the sociotechnical system (STS) (Reason, 2008; Woods et al., 2010; Amalberti, 2013; Hollnagel, 2014). A Safety Management System (SMS) describes the organisational arrangements to identify and apply safety controls in the workplace (Leveson, 2011). When combined with Morel et al.'s (2008) total safety approach that seeks to control risk generated by hazards present in the STS, Total Safety Management (TSM) emerges. TSM extends the SMS construct to account for the entire network of systems within systems comprising 'as designed' and 'as done' controls to achieve the required level of safety resilience in the workplace (Hollnagel, 2014; Leva et al., 2015). TSM, therefore, signposts a wider systems approach to organisational safety resilience that reaches across all aspects of the operating environment to optimise the overall performance of the organisation in achieving its safety aims (Cooper & Phillips, 1995). Consequently, this can produce wider benefits such as improved productivity, which is discussed in the next chapter, when considering the cost versus benefit of Individual Latent Error Detection (I-LED) interventions integrated within a TSM approach. This chapter reviews the elements shown in Figure 8.1, which proposes a TSM construct for delivering system safety, against which it is argued that optimising safety controls within the construct shown constitutes organisational resilience. This includes the integration of I-LED interventions such as those described in Chapter 7, against which it will be argued that the success of a TSM approach for safety resilience is predicated on competent operators who are the front-line users of I-LED interventions during everyday normal operations.

8.2 TSM CONSTRUCT FOR ORGANISATIONAL RESILIENCE

Human error effects are inevitable and occur daily (Reason, 1990; Hollnagel, 1993; Maurino et al., 1995; Amalberti, 2001; Perrow, 1999; Wiegmann & Shappell, 2003; Woods et al., 2010). Safety resilience requires effective system controls applied to the operating environment but also encompasses the human ability to adapt and overcome safety-related disturbances in complex STSs, which includes limiting the inevitability human-error effects due to system-induced performance variability (Hollnagel et al.,

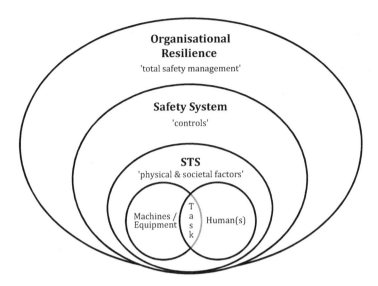

FIGURE 8.1   Total safety management construct depicting organisational resilience.

2006; Woods et al., 2010). Chapter 2 highlighted that the naval aircraft engineer maintains aircraft and equipment in the dynamic and complex STS for military operations, which gives rise to multiple safety-related hazards. Here, the engineer operates in multiple environments such as the maintenance office where aircraft documentation is completed and tasks planned; the maintenance hangar; stores for parts; issue centre to collect tools; and the aircraft operating line (ramp) or ship's flight deck to launch, turn around and also service aircraft, which is compounded further by time pressures, extremes of weather, constantly changing requirements due to emergent work or changes to the flying programme and resource constraints in terms of equipment, spares and people; operating aircraft from temporary airfields with very limited resources; operating from a moving platform whilst embarked in a warship; working on armed aircraft; and significant operational imperatives. Figure 8.2 provides context through a typical extreme cold weather operating environment, with further examples given in Appendix A.

It is widely recognised that resilience comes from progressive safety strategies that address system deficiencies through the identification and control of the network of hazards existing in the STS that can cause human failures leading to harm (Hutchins, 1995; Hollnagel et al., 2006; Reason, 2008; Woods et al., 2010; Leveson, 2011; Cornelissen et al., 2013; Dekker, 2014; Hollnagel, 2014; Chiu & Hsieh, 2016; Saward & Stanton, 2017). A resilient system, therefore, can detect and recover from safety failures caused by system hazards before they can cause harm. A key function of I-LED is its contribution to resilience in aircraft engineers by helping to mitigate for the inevitability of error across the full range of operating contexts and human performance factors. In the case of I-LED, Reason (2008) viewed humans as 'heroes' where behaviour exists that adapts to system failures to produce a safe recovery, which supports resilience. Similarly, Hollnagel's (2014) modelling of accident causation highlighted Safety II events

FIGURE 8.2 Naval air engineers working in an extreme cold weather environment. (Picture Crown Copyright ©.)

where the adaptive capability of human operators can locally overcome or avoid system failures, whilst his Safety I analogy refers to error avoidance and capture through the planning and delivery of effective safety controls aimed at defending against identified hazards. I-LED is a Safety II example where system cues trigger recall of past errors upon which a 'heroic' recovery can be made. Thus, it is believed that the safety aim of an organisation should not be preventing all errors from occurring, as it is arguably impossible to identify the entire network of potential system hazards and exceptional circumstances that can cause harm. The aim should be to maximise safety resilience as an enabler to achieving the required safety performance by mitigating performance variability induced by gaps or weaknesses in system controls. The following sections argue organisational resilience comes from a TSM approach that recognises the need to optimise system controls with safe behaviour associated with competent operators to help ensure successful latent error detection occurs before a causal path leads to harm.

8.2.1 Human Factors Integration within the STS

Amalberti (2013) highlighted that there are few models describing a framework for a global approach to the management of safety other than that generated at the local level to meet the specific safety needs of an organisation. Harris and Harris (2004) offer their '5M's' model to describe the complex sociotechnical relationships between worker, equipment and the organisation. At the centre of the 5M's model is the Human Factors Integration (HFI) between the (hu)**M**an-**M**ission-**M**achine interfaces, where the 'Mission' mission' refers to the 'Task' shown in Figure 8.1. The broad science of Human Factors and Ergonomics (HFE) considers the network of sociotechnical interactions between these interfaces comprising humans and technical aspects of the system, including machines, technology and processes (Edwards, 1972; Reason & Hobbs, 2003; Woo & Vincente, 2003; Carayon, 2006; Walker et al., 2008; Amalberti, 2013; Wilson, 2014; Niskanen et al., 2016). These networks can be complex in aircraft maintenance in terms of the number of interactions between systemic factors such as tools, equipment, procedures, organisational decision making, operator training and experience (Edwards, 1972; Reason, 1990; Reason & Hobbs, 2003). A Management layer encapsulates these HFI interfaces, which recognises the need to manage safety controls within a particular STS, that is the 'safety system' element shown in Figure 8.1. Harris and Harris (2004) also highlighted the dichotomy between physical and societal Mediums where the safety system must control factors influencing 'what can be done' within the organisation's physical medium against 'what should be done' as judged or directed by wider socio-political factors within the societal medium. In safety terms, it is argued socio-political factors will drive the level of safety performance needed by the organisation to meet regulatory and legal requirements (international, national and Defence), revenue projection, productivity, cost of litigation (cost of a safety failure), cultural influences and public perception of the organisation's safety credentials. This is especially true for safety critical industries such as construction, oil and gas, nuclear, commercial aviation, medical and transportation (Hendrick, 2003; Stanton and Baber, 1996; Goggins et al., 2008; Amalberti, 2013). Arguably, safety equilibrium is achieved when the organisation reaches consensus on what should be done against what can be done within the physical operating environment, which encompasses the operating environment, equipment design, training of competent operators, procedures and so on. Expectedly, military organisations manage safety against similar societal considerations, albeit revenue is replaced with the delivery of safe and cost-effective capability. Harris and Harris (2004) 5M's 'what can be done' in the physical medium can be further delineated through Hollnagel's (2014) 'as designed' controls and 'as done' safety strategy, for which the latter relates to safe behaviours in the workplace. I-LED interventions are dependent on system controls such as training, formalised procedures and, critically, the availability of cues to trigger recall (Saward & Stanton, 2017). Like the perceptual cycle model (PCM) described in Chapter 5, the effectiveness of the TSM model shown in Figure 8.1 relies upon bottom-up (BU)/top-down (TD) matching of real-world safety behaviour in the workplace with 'as designed' safety controls. This requires clearly defined contexts and limitations comprising 'what can be done' in the operating environment and assurance of 'as done' safety behaviour associated with competent individuals and teams (Morel et al., 2008; Harris & Harris, 2004; Amalberti, 2013; Hollnagel, 2014; Stanton & Harvey, 2017).

Deficiencies in an organisation's system controls can lead to uncontrolled hazards that transition to safety failures. Organisational accidents occur when there is insufficient Situational Awareness (SA) of the hazards that cause system failures, and/or there is ineffective control of the interacting component parts of the STS (Leveson, 2004). Thus, safety is created through effective risk management of hazards (Amalberti, 2013; Saward & Stanton, 2017), for which Morel et al. (2008) argued total safety is the product of controlling safety risks within the physical medium (such as rules and procedures, training and experience, supervisory controls, etc.) and managing risk-based activity locally through the adaptive abilities of competent operators. This is an important function of the proposed TSM model, as it takes Morel et al.'s (2008) total safety strategy to complement Leveson's (2011) focus on system controls alongside the dichotomous strategies offered by Hollnagel's (2014) causation model (Safety I and II approaches) and Harris and Harris (2004) view of the STS medium (what should be done versus what can be done). What can be done and controlled in the physical medium raises a further dichotomous situation where a resilient organisation must work hard to close the gap between 'as designed' controls and the real-world 'as done' safety behaviour in the operating environment. Here, I-LED promotes the operator's ability to self-monitor for system hazards and correct as necessary to help manage safety at a local level in the workplace, thereby contributing to resilience. Arguably, this function of I-LED facilitates 'as done' Safety II behaviour in the workplace that helps counter safety failures combing to create a causal path to harm (Stanton & Baber, 1996; Reason & Hobbs, 2003; Hollnagel et al., 2006; Woods & Hollnagel, 2006; Woods et al., 2010; Reiman, 2011; Cornelissen et al., 2013). I-LED interventions need to be integral to the safety system to be effective, which was highlighted in Chapter 7. Therefore, I-LED is also argued to contribute to resilience through Safety I strategies, provided it is integrated fully within each element of the TSM. The following explores each element of the TSM construct in more detail.

8.2.2 SAFETY SYSTEM

The safety system needs to articulate the strategy by which safety is to be managed within an organisation (Johnson & Avers, 2012), although Kleiner et al. (2015) noted that there is no globally agreed format for a management strategy, which supports Amalberti's (2013) comments earlier. At the organisational level, Leveson (2011) offered an SMS must describe how safety hazards are mitigated through structured safety controls appropriate to each level of responsibility in an organisation alongside a safety policy within the SMS that clearly defines safety boundaries with other organisations. Leveson also recognised the need to populate the workplace with competent operators who must be appropriately trained and risk aware. These are similar competence attributes reported by Flin et al. (2008), which is considered later. Thus, the safety system element shown in Figure 8.1 attempts to highlight the layers across which safety risks are identified and controlled within an organisation. The STS element is important as it provides the context and limitations for hazards across the network of HFI interfaces in the STS (Leveson, 2011), that is the safety system must be a good fit within the STS so that the 'as designed' and 'as done' elements are matched to facilitate maximum HFI safety performance (Hollnagel, 2014).

I-LED events improve safety by supporting resilience through Safety II events associated with the detection of latent error conditions. It is argued that a safety intervention is unlikely to be successful if not theory driven and matched to the requirements of the organisation's safety aims and objectives. Aligning to Leveson's (2011) SMS, this supports continual improvements in safety capabilities but requires management buy-in and leadership, changes to policy and training/education. The most effective safety interventions are, therefore, integrated with the enduring safety goals or strategy of an organisation rather than short-term activities such as occasional training sessions, and operators should be actively involved with interventions as this drives safety behaviour (Mullan et al., 2015). I-LED research in Chapter 6 described a golden window of two hours in which most errors were recalled. Simply waiting for schema housekeeping to occur within a time window of two hours is an example of passive intervention if there are no other safety controls in place, or existing controls are ineffective. This intervention becomes an active intervention or system control if the maintenance organisation mandates a two-hour break after all maintenance is completed before the aircraft is authorised for flight. I-LED interventions comprising the word and picture booklets plus Stop, Look and Listen (SLL) are all examples of active interventions. Applying Mullan et al.'s (2015) view on safety interventions, it can be argued that the active application I-LED intervention is most likely to deliver greater safety benefit than simply waiting for chance recall within the golden window of two hours. This provides further agreement that I-LED interventions should be integrated throughout the safety system to form a long-term strategy in support of resilience against system-induced latent error conditions.

8.2.3 ORGANISATIONAL RESILIENCE

A safety system involving complex activity should focus on resilience to counter the human effects from performance variability as well as the emergence of unexpected safety disturbances (Woods et al., 2010). This is important for safety critical organisations that need to be safe or even ultra-safe where one disastrous accident per 10 million events is aspired to (Amalberti, 2001), for example in aircraft maintenance. To aspire to greater resilience, the organisation should recognise that every element in the network of sociotechnical interactions contributes to the organisation's safety goals, where networks of multiple hazards exist that must be identified and controlled to create safety (Leveson, 2011; Plant & Stanton, 2016). A key function of the TSM construct is its resilience to safety disturbances within the STS, which suggests a TSM approach is appropriate to help the overcome safety-related challenges associated with the typical aircraft maintenance environments described earlier. The TSM construct in Figure 8.1 depicts a generalised framework for resilience in a safety critical organisation that is founded on competent operators, that is the 'human' component in the Harris and Harris (2004) 5M's model. However, it is recognised that the generalised construct belies the multiple interacting subsets and complex communication paths within a network of systems in systems that constitute the living and ever adapting world of sociotechnical networks (Hollnagel, 2014; Stanton & Harvey, 2017), that is networks comprising political, regulation, technical, economic, educational and cultural influences on human performance that directly impact the effectiveness of control measures embedded within the TSM construct (Kleiner et al., 2015). Resilience also reflects the need to optimise control measures at the organisational level through to the competence of individual operators in the workplace (Reason, 2008; Woods et al., 2010; Amalberti, 2013; Hollnagel, 2014).

8.3   OPERATOR COMPETENCE

It has been argued that the TSM approach to making the safety system safer is predicated on the presence of competent operators in the workplace, that is the heroes (Reason, 1990). This is not being critical of human error effects as the following section discusses, as it is the system that determines the level of competence required

(Hollnagel, 2014: 'as designed') and not the individual. The following section argues that additional mitigation for system-induced errors comes from enhancing the HFI element of the 5M's model by ensuring competent operators in the workplace, which is a Safety II function (Hollnagel, 2014). Competence characterises the expert operator and is especially important where safety controls are exhausted through exceptional conditions (Hollnagel et al., 2006; Reason, 2008; Amalberti, 2013; Chatzimichailidou et al., 2015; Saward & Stanton, 2017). This can include occasions where operators are exposed to system-induced hazards. For example, existing rules and procedures are found ineffective or unavailable for a specific task, equipment is poorly designed or not available, or other sociotechnical factors such as fatigue, task pressure and workplace distractions that can all lead to safety failures due to human error effects.

To help avoid safety failures in aircraft maintenance, as for most safety critical activities, the organisation can take an error-suppression approach to safety that relies on networks of safeguards and interlocks for controlling risks within the operating context, that is by maximising the use of documented procedures and exacting standards, reinforced with a quality management to help protect against system failures (Kontogiannis, 2011). Accident causation modelling continues to yield additional system controls to mitigate for potential causes of safety failures, which can include training, new procedures and more 'human friendly' machines and equipment. This increases error suppression, which might be counter productive from a systems perspective due to imposing excessive demands on human performance that may compromise the adaptive safety behaviours seen in competent operators (Amalberti, 2001). Amalberti also argued that over-optimising controls through processes and error-tolerant designs can reach a point where it is counter productive to improving safety, as the system needs to benefit from an element of flexibility to adapt to exceptional circumstances or uncontrolled system hazards that lead to error. This aligns with Rasmussen's (1997) concern that limiting the adaptive flexibility of operators overly constrains the safety system and arguably dilutes resilience through limiting Safety II events. Prescribed safety controls are essential in a resilient safety system (Leveson, 2011, Hollnagel, 2014), but, arguably, the ability to adapt to unidentified system hazards or unplanned exceptional circumstances also requires the cognitive skills to respond effectively rather than overly relying on error suppression techniques. Reason (1997) is sympathetic to the human condition and offered increasing the number of safety controls increases complexity that can create new opportunities for human error effects. As mentioned previously, Reason (2008) later offered the view of humans as heroes since they exhibit the ability to adapt to exceptional circumstances to detect and recover from their own errors. Indeed, humans continually adapt to their surroundings and modify behaviour in response to the often dynamic nature of an STS. Studies have shown that competent operators rely on system controls for safety critical activities but are also able to adapt their performance to manage recoveries from a significant amount of their own errors (Kanse, 2004; Thomas, 2004; Nikolic & Sarter, 2007; Flin et al., 2008; Malakis et al., 2010. Kontogiannis (2011) argued that the error suppression approach could be relaxed in favour of promoting error detection and correction in operators, for example through the use of I-LED interventions. Arguably, this highlights the need for a level of safety behaviour in operators, which is a product of system controls matched with competent operators, especially for the exceptional circumstances highlighted earlier.

Dekker (2014) reminds safety organisations that it is rare an individual goes to work to cause an accident (his 'bad apple' analogy), and thus, safety failures should be mitigated through system controls at the organisational level rather than relying on consistent human performance. But to design a safe environment in which humans operate successfully, you must have knowledge of human performance shaping factors – otherwise you do not know what deficiencies you are mitigating for and therefore the level of competence required in operators, that is what controls need to be designed and integrated within the safety system. Without knowledge of human behaviour, it is not likely system deficiencies can be mitigated through control measures that create resilience. For example, the analysis of diaries completed by naval air engineers in Chapter 7 found those with a high Cognitive Failures Questionnaire (CFQ) score are likely to be less receptive to external cues that trigger the appropriate schema response. Arguably, you cannot design effective system cues (accounting for system deficiencies) if you have insufficient knowledge of how receptive your general cohort are to certain cues. For pilots, the aircrew selection process facilitates mitigation for the safety expectations of the flight deck. When tackling safety deficiencies within this environment, it would be extremely challenging to design the flight deck to accommodate the full range of cognitive behaviours associated with all walks of life – unless your design goal is to remove the pilots completely. Arguably, if any human can fly a plane, operate a nuclear power plant, conduct medical operations or maintain complex aircraft, then all conceivable system deficiencies will need to mitigate for all extremes of potential variance in operator competence. Therefore, without a human-centred approach to error and safety, the STS view is perhaps counter intuitive, as you will end up removing the human from the system to achieve the safety aim (via autonomous machines). This approach to safety is likely to be too costly (in terms of financial costs), currently technically impossible for all operating environments and socially unacceptable. We may end up here in the very far future, but currently, society still needs to use all available tools to design safety controls to account for performance variability, which arguably includes Reason's (2008) heroic recoveries concept. Assuring operator competence is a safety control and is therefore argued to be essential for resilience.

A competent operator possesses the necessary error detection skills (awareness of the genotype/phenotype mismatch described in Chapter 5) to respond effectively to system cues. This includes a capacity for schema housekeeping to occur within the PCM, during which the perceptual cycle automatically reviews the effectiveness of BU/TD cognitive processing associated with past events (as opposed to detections that might occur due to chance). Here, I-LED interventions are argued to promote error detection post-task completion without risking error suppression, since all the interventions tested in Chapter 7 were 'process light' yet promoted SA regain within an operator's perceptual cycle. Thus, it is argued employing I-LED interventions improves operator competence to be ensured in the workplace, in pursuit of organisational resilience. The subject of organisational resilience is not a new concept, though (refer to Kleiner et al., 2015; Niskanen et al., 2016), but helping to ensure safe behaviours through enhanced operator competence, combined with I-LED interventions integrated within the TSM construct in Figure 8.1, is argued to be a new safety concept. Figure 8.3 provides a visual representation of the network of factors influencing operator competence, which are discussed in the following section.

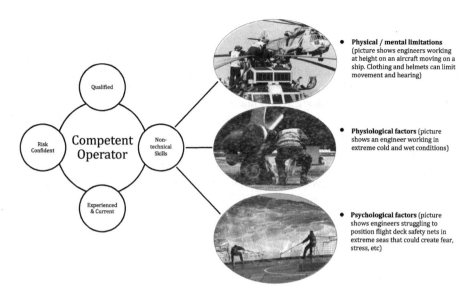

FIGURE 8.3 Factors influencing operator competence. (All pictures Crown Copyright ©.)

8.3.1 QUALIFIED AND EXPERIENCED

A resilient safety system needs to ensure operators are trained and possess the necessary skills to work safely and effectively in the operating environment (Flin et al., 2008). For example, a naval air engineer needs to be trained not only in aircraft maintenance but also in air operations, ship operations, logistics, quality assurance and military skills. This presents a significant training burden, which needs to also ensure the engineer is sufficiently experienced and current to carry out a task effectively. Importantly, experienced operators who have become experts exhibit the enhanced ability to detect and recover from their errors, which includes the detection of errors post-task completion (Wilkinson et al., 2011; Saward & Stanton, 2015b). Thus, qualification and experience are argued to be essential components of competence.

8.3.2 NON-TECHNICAL SKILLS

Competency is borne from safety controls comprising technical and non-technical skills (Flin et al., 2008). For aircraft engineers, technical skills typically include the use of tools and equipment, fault diagnosis, use of technical publications, following technical procedures, physical skill-of-hand and so on. Non-technical skills are the *cognitive, social and personal resource skills that complement technical skills* (Flin et al., 2008 p.1). The non-technical skills element of the competence model is expanded in Figure 8.3 and has been developed from a concept by Campbell and Bagshaw (2002), who offered human performance variability can be influenced by the Performance Shaping Factors (PSF) shown, which generally refers to sociotechnical factors that can increase occasions of human error or erroneous acts (Kirwan, 1998).

PSFs relate to an individual's response to system-induced factors associated with physical/mental limitations, physiological conditions and psychological conditions, all of which are argued to be characteristic of non-technical skills. These PSFs align with descriptions of non-technical skills, for which examples include (but are not limited to): physical/mental limitations – anthropometric reach, cognitive ability, vision, strength and so on; physiological conditions – extremes of weather, heat, vibration, noise, ship roll and so on; and psychological conditions – stress, fear, mental fatigue and so on. PSFs can influence safety behaviour (Kirwan, 1998), as the effects can manifest as reduced operator readiness (for a task). Particularly, studies have shown that PSFs mostly influence cognitive performance and, therefore, the error detection ability of the perceptual cycle, leading to insufficient SA (Gould et al., 2006; Liang et al., 2010; Plant & Stanton, 2013a). I-LED has been cited as offering an additional safety control to further mitigate for error events. Arguably, if the organisation does not control the influence on operator competence due to PSFs, then this might impact the ability to enact an I-LED intervention successfully.

Flin et al. (2008) also highlighted seven examples of non-technical skills: SA; decision making; communication; teamwork; leadership, managing stress; and coping with fatigue. These non-technical skills could all be considered PSFs. Particularly, I-LED events occur when ongoing schema housekeeping (Saward & Stanton, 2015a) or surrounding cues trigger a review of schemas applied to past tasks, that is the schema-action-world cycle. This cognitive processing of cues associated with the workplace is essential for SA regain. Naturalistic Decision Making (NDM) refers to occasions where people decide to act based upon relating prior experience to their perception of real world events and is therefore schema driven and can also be related to the PCM (Klein, 2008; Plant & Stanton, 2013a). This draws synergies with decisions made in response to schema selection and enactment based upon the perception of cues to trigger the decision-making process. Thus, it could be viewed that I-LED is an example of NDM. However, NDM is concerned with making decisions based upon informed choices, whilst I-LED is concerned with the recall of past errors when triggered by a related system cue. Thus, I-LED is believed to offer a distinct human quality, in addition to the non-technical skills defined by Flin et al. (2008).

8.3.3 Risk Confident

Whilst adjunct to the other elements of the competence model shown in Figure 8.3, being confident with risk management is ability predicated on an operator being qualified, experienced and personally ready for the task in hand. Without these preceding elements, arguably, the operator is not likely to recognise safety risk (borne from hazards present in the STS) in terms of where system failures might occur and how to respond effectively. Leveson (2011) highlighted the need for risk awareness throughout the operating environment. This could see the operator take action to avoid the system failure; to correct for the failure proximal to the error event or detect and recover from a latent error condition before it networks with other safety failures in the STS to form a causal path. Thus, it is argued risk confidence is the product of being risk aware and possessing the skills to respond effectively to a safety failure, which might be highlighted through an I-LED trigger.

8.4 SUMMARY

There appears to be few models describing an agreed safety strategy for achieving organisational safety resilience. Thus, a TSM construct has been proposed to highlight the hierarchical relationship between safety 'as done' (Safety II) by competent operators in the workplace through to the 'as designed' (Safety I) controls needed for system safety. It is recognised that the generalised hierarchy of TSM construct introduced in this chapter belies the complexity of real-world operating environments where multiple interacting subsets and complex communication paths within a network of systems in systems reflects the true living and ever changing nature of a STS. However, it has been argued that the TSM construct provides utility in a simple framework from which to understand the layers across which safety needs to be optimised to achieve resilience. It has been further argued safety resilience is dependent on the ability of the system to promote safe behaviours, which is predicated on the presence of competent operators in the workplace who possess the necessary technical and non-technical skills that help mitigate for system-induced human error effects. Operator competence, and therefore safe behaviour, depends on system controls, which is congruent with systems thinking. I-LED is a safety control that is a non-technical skill, which promotes SA regain without risking error suppression. To achieve the Safety II benefit, it has been argued that I-LED interventions need to be applied at the local level by competent operators who are qualified, experienced and current for a task, risk confident, and able to enact an I-LED intervention despite being immersed in PSFs. Chapter 5 highlighted HFE-designed I-LED interventions that engage with system cues are especially important for simple everyday habitual tasks carried out alone or for tasks perceived to be low risk where perhaps human error effects are most likely to pass undetected if there are deficiencies in the organisation's safety-related defences. Thus, when successfully integrated within the safety system, I-LED is believed to contribute to organisational safety resilience through improved operator competence. Enhanced safety resilience through I-LED interventions integrated within a TSM approach to system safety should be of benefit to any safety critical organisation seeking a progressive strategy that results in safety successes, for which the cost-versus-benefit arguments associated with this approach are considered in the following chapter.

9 Assessing the Benefits of I-LED Interventions

9.1 INTRODUCTION

The previous chapter offered that safety is created from effective risk management that controls hazards across the sociotechnical system (STS), at the organisational level through to individual operators in the workplace. I-LED interventions facilitate individual operator engagement with system cues to detect latent error conditions, which arguably improves operator competence in the form of an additional non-technical skill that acts as a system control. This supports risk management by helping to regain Situational Awareness (SA), leading to the detection of system-induced errors post-task completion (Naderpour et al., 2014; Chatzimichailidou et al., 2015; Niskanen et al., 2016), which offers a paradigm shift in safety thinking as it offers a new control strategy for an organisation seeking to make their safety system safer through enhanced resilience.

The introduction of any new safety strategy is likely to incur financial costs as well as other resource implications such as training, new procedures and time to conduct the intervention. Thus, the costs versus benefit of integrating I-LED interventions within an existing organisational safety strategy needs to be assessed. Assessing the benefit of a safety intervention can be problematical, though, not least as Weick (1987) recognised *safety as a dynamic non-event;* thus, you do not necessarily know a specific safety control avoided a safety failure or defeated a causal network that could have resulted in harm. This was evident during the literature review in Chapter 2, where only three occasions of I-LED events reported in the Air Safety Information Management System (ASIMS) database were found for naval aircraft maintenance. Here, it was argued the database might hide the true effectiveness of safety controls due to under-reporting of failures and non-reporting of safety successes. Therefore, it is difficult to calculate truly representative costs associated with the benefit of specific safety controls.

The following chapter assesses the benefit of I-LED interventions as an additional safety control versus the financial costs and wider resourcing issues. This contributes to Objective 5, which is to assess the benefit of integrating I-LED interventions with organisational safety strategies to enhance resilience. To create safety through effective risk management, the risk analysis process identifies hazards and their potential impact upon system safety, and therefore, the benefit of specific safety controls needs to be considered first. Thus, the ASIMS database introduced in Chapter 2 is accessed for examples where I-LED interventions might have countered the hazardous events reported in the database, before calculating the cost of introducing an I-LED intervention as a new control measure.

9.2 RISK ANALYSIS

Risk is the likelihood that an uncontrolled hazard will result in undesirable outcome, measured against the impact or severity of the hazardous outcome such as an accident. Where the outcome threatens life, this can be denoted as Risk to Life or RtL (JSP 892, 2016; HSE, 2017). UK naval aviation is conducted within complex sociotechnical environments, delivered around the globe: from land bases in the United Kingdom with full aircraft support facilities to deployed temporary airfields with very limited facilities, and from large multi-aircraft carriers to small single-aircraft ships. The naval aircraft engineer operates within these operating contexts where safety-related hazards are an inherent and reasonably foreseeable part of normal operations. Risk analysis helps ensure these hazards are understood and safety controls implemented to protect aircraft maintenance from undesirable outcomes.

System-induced human error effects are widely cited as the most significant contributing factor in safety failures that lead to accidents, for which error is both inevitable and a daily occurrence (Reason, 1990; Hollnagel, 1993; Maurino et al., 1995; Perrow, 1999; Wiegmann & Shappell, 2003; Flin et al., 2008; Woods et al., 2010). Amalberti and Wioland (1997) showed errors suffered by highly trained and competent operators (such as engineers) can be frequent but are either inconsequential or detected and corrected before leading to an undesirable outcome. However, it is also recognised that many error events pass undetected to become a latent error condition that poses a risk of harm if not detected later, either as a singular event or if networked with other safety failures to form a causal path (Graeber & Marx, 1993; Hollnagel, 1993; Maurino et al., 1995; Reason, 1997; Perrow, 1999; Reason & Hobbs, 2003; Flin et al., 2008; Lind, 2008; Woods et al., 2010; Aini & Fakhru'l-Razi, 2013). Saward and Jarvis (2007) sampled 4428 ASIMS occurrence reports across UK military aviation and found an average of 5.25 system failures exist per occurrence, with most of these occurrences recorded with an event severity of only negligible, that is no injury to personnel or damage to an aircraft was experienced as a consequence. Chapter 2 cited a study by Graeber and Marx (1993) that analysed occurrence reports from a major airline over three years. From 122 maintenance errors, they found omissions (56%), incorrect installations (30%) and wrong parts (8%). The Civil Aviation Authority (CAA, 2009) commissioned a study of UK civil aviation mandatory occurrence reports involving jet aircraft (years 1996–2006) to find that incorrect maintenance actions and incomplete maintenance contributed 53.1% and 20.7% to be present, respectively (total n = 3284 mandatory reports). Physical examples include loose objects left in the aircraft, fuel caps unsecured and cowlings or access panels unsecured (Latorella & Prabhu, 2000). Most of the examples posed no immediate consequences but arguably generated a latent error condition that could have networked with other system factors (safety failures and other latent conditions) to cause harm. For example, the engine cowl doors shown in Figure 2.1 were left unlatched on an Airbus A320-231, which resulted in the doors detaching in flight (AAIB, 2000). The aircraft returned safely, avoiding an accident, but if the doors impacted the aircraft fuselage or struck someone on the ground, then the consequences clearly would have been much more severe.

FIGURE 9.1 Crash of a Hawk T Mk1 shortly after take-off.

A search in ASIMS (2017) provided an example where maintenance error led to severe outcome when the engineers omitted to reconnect the flying controls after maintenance on a military jet aircraft. The error remained undetected and resulted in the loss of the pilot and aircraft, shown in Figure 9.1.

The hazard was flying controls can be left unconnected. Safety controls to mitigate for this maintenance error event encompass procedures, documentation, training of competent engineers and supervisor independent checks to avoid or detect the safety failure. If not detected via these typical control measures, the now latent error condition presents a system risk, for which additional safety controls such as pre-flight checks and a visual check of the flying controls before departure are examples of further measures designed into the safety system to detect the latent error condition. In this example, though, the error was not detected and contributed to a causal path where it became too late to make a safe recovery. The ASIMS search provided a further example where a securing nut and bolt assembly was fitted incorrectly on a helicopter tail rotor system, shown in Figure 9.2.

Again, the latent error condition remained undetected, creating a new risk with the potential for the aircraft to crash. In this example, though, the aircrew made a successful emergency landing by responding in time to the subsequent failure of the tail rotor during flight. Here, controls designed to counter risk escalation defeated a causal path to a more severe outcome. Arguably, an I-LED intervention might have been of benefit as an additional safety control in all the previous examples to help detect the latent error condition before risk escalation. The potential costs of integrating I-LED interventions as an additional safety controls to benefit a more resilient safety system are considered in the following section.

FIGURE 9.2   Disconnected flying control on Lynx aircraft.

9.3   CALCULATING THE BENEFIT OF SAFETY CONTROLS

The need to demonstrate the benefit of safety controls has long been recognised in human factors research where their effectiveness against cost in terms of safety, economic, productivity and socio-political factors needs to be calculated (Hendrick, 2003; Stanton & Baber, 1996; Goggins et al., 2008). Micheli and Cagno (2009) conducted a survey of small-to-medium organisations to find that safety was a priority, but 80% (n = 109) of the organisations struggled to implement safety interventions due to the lack of financial evidence. Cost Benefit Analysis (CBA) assesses the cost effectiveness of safety controls aimed at mitigating potential safety failures (Stanton & Barber, 2003; Goggins et al., 2008). Therefore, safety controls need to be selected carefully using a risk analysis process and be based on demonstrable theory shown to yield a reduction in injuries or death whilst also improving overall system performance without disproportionate impact on time or financial costs (Gilbert et al., 2007; Leveson, 2011). Mullan et al. (2015), in their analysis of safety interventions within the construction industry, reviewed several safety intervention studies to find that legislation alone was not effective (at preventing injuries in the workplace), but interventions that changed behaviours through the active involvement of operators based on theory were effective. Further, an intervention should be applied long term to become part of routine safety activity rather than single training event or safety campaign. Similarly, Leveson (2011) argued investing in safety long term benefits a reduction in injury/death rates, improves productivity and overall output, and other goals such as socio-political status. But new safety controls can lack appeal to the organisation if they are poorly designed or applied incorrectly to the workplace, and thus, the safety benefit is not likely to be justified. Critically, Mullan et al. (2015) noted that operators should be actively involved in the safety interventions rather than a working to a new management process, as this empowers safe behaviours. Johnson and Avers (2012) found that improving system safety through strategies that improve the operators' working environment also improved overall operator performance or heroic recoveries through non-technical attributes such as empowerment and job satisfaction (Flin et al., 2008; Reason, 2008; Woods et al., 2010). Reason and Hobbs (2003) argued that maintenance errors are mostly attributable to financial losses or an impact to productivity rather than being directly causal to injuries or deaths, simply because of the number of system controls present in a safety-focussed organisation. The ability of I-LED interventions to counter other potential consequences of latent error conditions hidden in the system might also benefit the wider aims of an organisation such as improved productivity, overall system performance, socioeconomic gains, and political and reputational value (Kleiner et al., 2015). Thus, carefully designed I-LED controls can offer real-world benefits to the local management of safety but must be theory driven, engage directly with operator cognitive skills, be appropriate to the context employed and integrated within the overall safety system to improve safety.

The mathematical calculation is relatively straightforward, but it is the compilation of sufficiently accurate data on financial costs that possesses the greatest challenge to CBA calculations or assessing the Return on Investment (ROI: Johnson and Avers, 2012). The UK Health and Safety Executive (HSE) injury figures (HSE, 2016) show in 2015/2016 the United Kingdom suffered 144 deaths and 72,702 injuries, and between 2013/2014 and 2015/2016, an estimated £4.8 billion in economic costs due to injury or death. The Federal Aviation Administration (FAA, 2017) and HSE both offer online tools for ROI/CBA analysis. The HSE (2017) CBA checklist considers the costs of integrating the safety intervention into normal operations as well as training and the enduring support infrastructure needed to maintain a safety control long term. The checklist is purposely designed to assess the financial benefit against what should be done versus what can be done proportionate to avoiding RtL and/or environmental damage, and not the wider benefits such as improved productivity, damage to equipment, economic or socio-political gains. Johnson and Avers (2012) noted, though, that it is extremely challenging to calculate accurate and meaningful financial costs due to the complexity of the STS and expertise needed on factors such as equipment costs, training, cost of life or injury, and impact on lost output. Both ROI/CBA tools recognise the dependency on accurate cost data to model a valid cost versus benefit argument or percentage return on the investment, especially as many benefits offer the intangible gains discussed. Indeed, the realisation of an organisational accident from a causal path of undetected safety failures can come as a complete surprise to a safe or ultra-safe organisation that already invests heavily in their safety system (Amalberti, 2013). Here, it can be argued that if a new safety control is proportionate to the population that could reasonably benefit from reduced exposure to RtL, and it is not cost prohibitive, then the safety control should be introduced since not doing so could weaken safety resilience (Johnson & Avers, 2012; Amalberti, 2013; HSE, 2017).

Despite the challenges of CBA/ROI analysis, Objective 5 for the current research seeks to assess the benefit of I-LED interventions that have been argued to help enhance safety resilience. Thus, the following section analyses the RtL associated with typical maintenance-related error events reported in ASIMS, from which representative CBA and ROI calculations are provided for the introduction of an I-LED intervention tested in Chapter 7.

9.4   ANALYSIS OF I-LED INTERVENTIONS

9.4.1   METHOD

The same ASIMS database interrogated in Chapter 1 was accessed for all Royal Navy (RN) aircraft maintenance-specific safety reports from aircraft squadrons over the period 31 March 2012 to 1 April 2017 (ASIMS, 2017). The population of naval aircraft engineers employed in aircraft squadrons was approximately 1700 during this period, and five years was sampled to provide a sufficient number of reports that could be analysed within the resources available for the current research. The search returned 1571 reports over this period, which included: technical factors such as failed or worn components; data integrity issues; design issues; ineffective equipment or procedures; environmental factors such as adverse weather, erosion or corrosion issues and the impact of operating conditions on aircraft maintenance; and Performance Shaping Factors (PSF) (Kirwan, 1998). PSFs were described in Chapter 8 and encompass physiological, psychological and physical limitations that influence human performance leading to error effects. The reports were filtered for human performance factors only to align with the current research aims. This returned 627 reports for analysis.

The narrative from each report was reviewed for latent error examples (as opposed to system failures detected proximal to the reported event) and further filtered for reports where an I-LED intervention might have been of benefit. To achieve this filtering process, knowledge and experience of naval aircraft engineering and aircraft types were essential to interpret the technical narratives contained in the ASIMS reports (Schluter et al., 2008). To maintain the quality of the analysis, if a report contained insufficient data about the error event to claim that an I-LED intervention might have been effective, the report was discarded to avoid biasing the analysis. The lead author had full access to the database as part of his normal employment as an Air Engineer Officer (AEO). No identifiable or protected information (personal details, locations, aircraft types and equipment serial numbers) were accessed from ASIMS during the data mining, and the narratives were analysed within a Ministry of Defence (MoD) restricted Information Technology (IT) network. To test for inter-rater agreement, another AEO was used as an independent assessor to conduct a 100% review of the 627 reports, which included agreement on the potential to benefit from an I-LED intervention and the categories for 'latent error condition' and 'risk' shown in Table 9.1. Cohen's kappa was calculated on the frequencies shown as opposed to percentage agreement to correct for any chance agreement (Robson, 2011). This found k = 0.88, indicating very good agreement on which reports an latent error detection (LED) intervention is likely to have been of benefit.

Data mining revealed 40% (n = 249) of the 627 filtered reports to have the potential to benefit from one of the individual LED interventions tested in Chapter 7. Table 9.1 summarises narratives of typical error events recorded in ASIMS (bracketed text explains technical terms, and 'xxxx' is included where identifying information has been redacted). The latent error condition is shown along with an estimate of the worst case perceived risk if the condition had not been detected. The narratives have been grouped and counted according to the ASIMS perceived safety severity (ASIMS, 2013b), noting that an I-LED event is argued to provide mitigation for each severity

TABLE 9.1

Example Narratives from ASIMS Reports

ASIMS Severity	Example Narratives	Latent Error Condition	Worse Case Perceived Risk
A-High (n = 14)	'The APU (auxiliary power unit) fire bottle cartridge was found electrically disconnected during the AFS (after flight servicing)'.	Fire protection inoperative	Uncontrolled fire leading to loss of aircraft and/or life
	'I asked one of the AETs (air engineering technicians) to remove the cable cutter cartridges (from an aircraft rescue hoist). The AET tasked with removing the cartridges then informed me that there were no cartridges fitted. I went over to the hoist and confirmed there were no cartridges in place'.	Cable cutter inoperative	Aircraft unable to break-free if cable snagged leading to loss of aircraft and/or life
	'During entry into right-hand side front seat ... it was noticed that the holding open strut was upside down on the door. With the jettison handle fully forward, it was impossible to move the door out of the airframe (for an emergency escape)'.	Pilot egress route blocked	Loss of life in an emergency
B-Medium (n = 80)	'During task to replace the wire locking with split pins, post successful test fight, it was noted that the yellow main rotor blade pitch change link had been incorrectly built'.	Main rotor system installed incorrectly	Significant vibration on start and potential damage
	'During EGR (engaged ground-run) ... accessory GB (gearbox) inlet and exhaust cooling duct grilles found to be blanked with black masking tape'.	No gearbox cooling	Overheating leading to aircraft emergency
	'On completion of a period of flying oil was observed leaking from the transmission bay port and starboard common overboard drains ... the MRGB (main rotor gearbox) oil filler cap was found not fitted.'	MRGB oil uncontained	Oil depletion leading to aircraft emergency

(Continued)

TABLE 9.1 (*Continued*)
Example Narratives from ASIMS Reports

ASIMS Severity	Example Narratives	Latent Error Condition	Worse Case Perceived Risk
	'Myself and LAET xxxx were tasked with fitting xxxx wedges and functional test on the aircraft Chaff & Flare system. Each of the 2 wedges required 6 bolts, but we only had 8 available. We fitted these (4 on each) so LAET xxxx could continue with the functional testing while I tried to obtain the other bolts through main stores as we had no other available manpower at the time, and there was no squadron stores personnel working. After lengthy searching, I discovered that the NSN we had for the bolts was no longer valid, and I would have to talk to MODS Control to find an alternative number … but nobody at MODS Control. On returning to the aircraft, the M147 System had failed its functional test. I explained the situation with the bolts to LAET xxxx, and we decided to ensure the system is serviceable before trying to locate the remaining 4 bolts. Around 18:00, we completed a functional test on the xxxx system and proceeded with returning the tools and completing the aircraft documentation (making the aircraft serviceable for flight). On my way to work this morning at xxxx, I realised our error'.	Chaff and flare system installed incorrectly	Fuselage damage and/or failure of chaff and flare system
C-Low (n = 135)	'During an MTF (maintenance test flight) walk-round, the flying maintainer spotted that all five pitch change rod upper bolts were orientated incorrectly'.	Tail rotor system installed incorrectly	Vibration/minor damage on aircraft start
	'On xxxx, I was tasked with moving aircraft xxxx. After moving about 2 metres, I noticed something move on the starboard side and stopped the move … the bonding lead had snapped as it had still been attached to the aircraft'.	Bonding lead not removed	Fuselage damage
	'During the walk round prior to a PTF (partial test flight), it was noted that the starboard hydraulic oil filler cap and access panel had been left open'.	Hydraulic oil free to escape	Damage and/or emergency landing
	'Whilst preparing for engineering rounds, a Supervisor reported a spanner had been found resting on the Detachment Hydraulic Rig located just outside of the front of the Hangar'.	Metal object on aircraft operating area	
D-Negligible (n = 20)	'The Aaircraft Commander opened xxxx MF700 (aircraft documentation) and found an open entry. The engineering line was informed, and they removed the book coordinate correctly'.	Full serviceability of aircraft not known	D-negligible risk not reported as no reasonably foreseeable consequence
	'Upon start-up of xxxx for an EGR (engaged ground-run), the blade fold display indicated that the No. 2 rotary actuator was showing as a yellow. On investigation, the electrical connector for the No. 2 rotary actuator was found disconnected'.	Blade fold inoperative	

type. Additionally, the first-party Valuation of Prevented Fatality (VPF) amount has been quoted against each severity using HSE (2017) data, which includes a valuation for injury:

- *A-High*: There are few or no remaining safety controls that could credibly have prevented a loss of life or significant injury, leaving consequence to chance (VPF = £1,336,800 – single death).
- *B-Medium*: The safety controls are weak or can be missed, leaving a clear path to loss of life or significant injury (VPF = £772,000 – averaged single death and permanently incapacitating injury).
- *C-Low*: The safety controls appear adequate in the protection they offer against loss of life or significant injury (VPF = £20,500 – serious injury).
- *D-Negligible*: There is no readily conceivable means through which this occurrence could have led to a loss of life or significant injury (VPF = £530 – minor injury requiring one week off work).

9.4.2 Findings

ASIMS data recorded 627 maintenance events relating to human performance, which is 125.4 error events or 0.07 per engineer per year. This is an extremely low number considering human error occurs daily (Maurino et al., 1995; Perrow, 1999; Reason & Hobbs, 2003; Wiegmann & Shappell, 2003; Flin et al., 2008; Woods et al., 2010). This perhaps confirms that naval aircraft maintenance activity is already very safe or that maintenance error is under-reported or passes undetected to create a hidden latent error condition. Indeed, applying Bird's (1969) theoretical safety triangle, there are likely to be many thousands of unreported safety-related events that occur in large organisations. Thus, the absence of safety data compounds the challenge to demonstrably cost the true benefit of safety controls such as I-LED interventions. All examples in Table 9.1 are typical maintenance error events (Rasmussen, 1997; Amalberti, 2001; Reason & Hobbs, 2003), which arguably highlights the need for broadly applicable system controls designed to counter the typical hazards shown in Table 9.1, as it will be challenging to design theory-driven bespoke safety controls for specific events.

Analysis of ASIMS found 14 potential I-LED reports recorded with a high severity. The examples in Table 9.1 show that the worst-case RtL through fire, crash landing and/or the inability to escape in an emergency. Since each aircraft involved in the 14 maintenance-related reports is crewed with a minimum of two, there is a potential for loss of life or major injury to 28 crewmembers. The complexity of costing military personnel and equipment damage is beyond the scope of the current research, but using the HSE (2017) VPF figures highlighted earlier, representative CBA and ROI calculations for the Stop, Look and Listen (SLL) intervention described in Chapter 7 are provided in the following section.

9.4.3 Estimating the Cost of an I-LED Intervention

To illustrate the potential cost versus benefit of integrating an I-LED intervention within the aircraft maintenance safety system, the SLL intervention described in

Chapter 7 can be costed in general terms. This intervention has been selected as it was found to be the most effective safety control when compared to the other interventions tested. The approximate average capitation rate (cost to Defence, as opposed to salary) for a naval aircraft engineer is approximately £62,000 and employed in the aircraft maintenance environment around 210 days a year (allowing for training courses and extraneous military duties not involving aircraft maintenance). Thus, the approximate cost to Defence per day is £295.24, from which a maintenance man-hour (MxMHr) = £36.90 (allowing for an 8-hour working day). An aircraft engineer is estimated to work on 10 maintenance tasks per day, for which the SLL intervention will take 2 minutes to enact at a cost of £12.30 per working day (MxMHr × 20 mins). The SLL intervention was found to be quick to apply and did not require any additional material such as a booklet, printed procedure or equipment. Thus, the cost of this new safety control is argued to be simply the cost of initial training plus the time to apply the SLL intervention to each maintenance task per working day. Thus, the cost of the SLL intervention for one air engineer over 1 year is £2,620 (initial training £36.90 + £12.30 × 210 days, and assuming refresher training will be included within existing annual human factors training). Chapters 4 and 5 highlighted the population studied in the current safety research consists of around 1700 naval aircraft engineers employed in naval aircraft helicopter squadrons. Thus, the total Cost of Integration (CoI) in this population is £4,454,000 in the first year. The CoI is the cost of integrating the SLL intervention throughout the safety system across the four event severities recorded in ASIMS. Thus, the CoI per event severity is £1,113,500. Table 9.2 has been constructed with the CoI, from which illustrative CBA and ROI have be calculated, noting that the analysis does not include additional costs such as equipment damage, third-party harm to people or any impact to the environment.

Figures for CBA and ROI can be calculated as follows (Stanton & Young, 1999; Johnson & Avers, 2012; HSE, 2017):

- CBA (£) = [(RtL) × (detection performance, Φ)] − (CoI)
- ROI (%) = CBA/CoI

CBA and ROI calculations shown in Table 9.2 suggest the most significant financial benefit comes from the SLL intervention applied to medium-severity events due to the higher RtL value. High-severity events also suggest significant financial benefit. High and medium events were anticipated to show significant CBA and ROI can be achieved since a credible and reasonably foreseeable RtL exists in the maintenance error events recorded in ASIMS.

The majority of the 249 potential I-LED occasions found in ASIMS where recorded with a low or negligible severity, which supports Amalberti and Wioland's (1997) finding that errors made by highly trained and competent operators are either inconsequential or detected and corrected before leading to an undesired outcome. Notably, Chapter 4 described some risks associated with uncontrolled hazards can cause a system failure but pose little impact to overall system safety if not detected. The calculated values in Table 9.2 appear to support this view in that the low-severity events showed a small financial benefit compared to high and medium events, whilst

TABLE 9.2
SLL Intervention Cost Calculations (Represented over One Year)

No. Events (Table 9.1)	Severity (Table 9.1)	Likelihood (Events per Year)	VPF (£)	RtL Value (£)	Phi, Φ	CoI (£)	CBA (£)	ROI (%)
14	High	2.8	1,336,800	3,743,040	0.67	1,113,500	1,394,337	125
80	Medium	16	772,000	12,352,000	0.67	1,113,500	7,162,340	643
135	Low	27	20,500	2,767,500	0.67	1,113,500	740,725	66.5
20	Negligible	4	530	10,600	0.67	1,113,500	−1,106,398	–

Note: Likelihood – Average number of ASIMS events per year.

VPF – HSE (2017) valuation of prevented fatality or injury.

RtL Value – Valuation based upon average number of event severity per year (likelihood) × VPF.

Φ – Phi (Φ: Matthews, 1975) for the SLL intervention is given in Table 7.3 (average score for operatives and supervisor combined).

CoI – Cost of integrating the SLL intervention.

the negligible category reported a negative CBA and, therefore, no ROI against the potential RtL shown. It has been highlighted that an organisational accident is rarely the result of one error effect, since it is more usually the confluence of more than one safety failure that creates a causal path to an accident (Amalberti, 2013). Also, the study in Chapter 6 reported that habitual tasks carried out alone or tasks perceived by the individual to be low risk were at particular risk of human performance variability. Thus, it is believed that even the low- or negligible-event occurrences recorded in ASIMS could also benefit from an additional safety control such as an I-LED intervention due to the risk of latent errors networking with other factors to form a causal path.

The SLL intervention is the simplest I-LED intervention of those tested in Chapter 7. Although the information and the subjectivity of the narratives from ASIMS meant that it was difficult to decide with confidence which specific LED interventions offers the greatest safety benefit, it is argued that the SLL intervention benefit also comes from its broad applicability across various maintenance activity. The other three interventions require additional resources to be considered in the cost calculations. For example, the booklets containing words and pictures need to be produced in sufficient quantities to be widely available on an aircraft squadron, which averages around 120 engineers per squadron. These booklets also need to be managed carefully to ensure they are used effectively and do not in themselves become a loose article left on an aircraft, thereby rendering the error countermeasure redundant. This presents a practicable issue to ensure all engineers have access to the booklets at any time and in any location or operating environment. This can be managed in a UK base but rapidly becomes impracticable when maintenance is deployed in the field. There will also be an ongoing cost to ensure the booklets are regularly updated to reduce the chance of engineers becoming desensitised to the words and/or pictures. It is thought that the management of these challenges is likely to reduce the benefit of the booklets above and beyond existing LED safety controls such as supervision or procedural checks mentioned earlier.

9.5   SUMMARY

This chapter has assessed the benefits of introducing I-LED interventions as an additional safety control within aircraft maintenance to enhance resilience, which addresses Objective 5. The benefit of an I-LED intervention comes from its integration with an organisation's safety system to form part of an enduring long-term safety strategy to engender and control safety behaviours in the workplace (Mullan et al., 2015). Analysis of ASIMS data revealed occasions where latent error events might have been prevented if an I-LED intervention had been used as an additional safety control. Risk analysis against VPF figures suggest significant CBA and ROI can be achieved where the perceived severity of the reported safety failure is high or medium. Here, a credible and reasonably foreseeable RtL existed in the maintenance error event reported in ASIMS. The representative calculations also showed a small CBA/ROI argument for low-severity events and a negative return for negligible events.

It has been highlighted that little is known about how networks of latent error conditions might combine with other latent conditions and safety failures to form a causal path to an accident, especially for everyday routine and perhaps less safety critical maintenance tasks such as those reported in Chapter 6. Since organisational accidents are rarely the result of a single safety failure (Amalberti, 2013), combinations of more than one failure, irrespective of its individual perceived severity, can cause the overall RtL to escalate to cause harm. Thus, it has been argued safety controls are essential in a resilient safety system and should be applied universally across all types of safety activity. The integration of I-LED interventions as additional safety control has also been argued to offer additional benefits in terms of reduced equipment damage and economic gains as well as intangible socio-political effects and non-technical attributes such as improved operator empowerment and job satisfaction.

10 Conclusions and Future Work

10.1 INTRODUCTION

The aim of this book has been to explore the nature and extent of Individual Latent Error Detection (I-LED) in aircraft maintenance and its contribution to making the safety system safer through enhanced resilience. Safety research has been framed around the objectives stated in Chapter 1 along with three linked observational studies to investigate the I-LED phenomenon in naval aircraft engineers in their normal everyday working environment during everyday operations. A summary of this new safety field is provided in the following sections, along with an evaluation of the approach to research and directions for future work. The concluding remarks offer a final statement of how I-LED interventions should be integrated within the system safety to enhance safety resilience in naval aircraft maintenance, as well as wider safety critical organisations.

10.2 SUMMARY OF SAFETY RESEARCH

10.2.1 OBJECTIVE 1

Using a human-centred systems approach, develop a theoretical framework to observe the I-LED phenomenon.

To address the first objective, an extensive review of literature surrounding the proposed I-LED phenomenon was conducted in Chapter 2. This highlighted system-induced human-error effects to be the most significant factor impacting the safety success of an organisation. Human error is inevitable and occurs daily, for which undetected error becomes a latent condition within the sociotechnical system (STS) that can network with other safety failures to create a causal path to an accident if not detected. Whilst system causes of human error have been researched widely, as has error avoidance and proximal detection, no specific research was found on I-LED. This confirmed the phenomenon to be a novel concept, indicating a new safety field requiring research. In the absence of specific research, a multi-process theoretical framework was developed from existing theories on Prospective Memory (PM), Supervisory Attentional System (SAS) and schema theory. To determine whether the theoretical framework aligned with real-world evidence, thematic analysis of Aviation Safety Information Management System (ASIMS) data found the phenomenon existed in naval aircraft engineers and that the multi-process framework appeared to be suitable theory, against which to conduct I-LED research.

The initial literature review also revealed some disquiet exists over the use of the term human error, as it can be used to blame individuals rather than signpost the

119

opportunity to tackle deficiencies within the STS. Progressive safety strategy requires a systems view but equally must not forget that the individual human is at the sharp end of the safety solution where system-induced performance variability can lead to latent error conditions. Multi-process theory seeks macro ergonomic solutions to achieve I-LED; thus, use of the term human error was argued to be congruent with systems thinking and meaningful in progressive safety research, provided the term is used carefully from a systems perspective to describe performance variability effects caused by the full range of system influences, rather than just focussing on individual human failures.

Chapter 4 highlighted organisational accidents are rarely the result of a single cause; thus, in terms of undetected errors, there is often the confluence of more than one latent error condition and other safety failures that impact system safety. The I-LED phenomenon introduced in Chapter 2 has been argued to act as an additional safety control that supports the detection of past errors by the individual who suffered the error using system cues to trigger recall, without which the latent error condition can network with other safety failures to form a causal path of escalating risk. I-LED interventions can contribute to existing hazard controls such as process checks and independent inspections to enhance safety resilience, for which the number of controls or interlocks is a function of the level of safety required by the organisation to achieve its safety aim.

10.2.2 OBJECTIVE 2

Apply the theoretical framework to understand the nature and extent of I-LED events in naval aircraft engineers working in their everyday maintenance environment.

In Chapter 5, data were collected from a cohort of naval aircraft engineers via a group administered questionnaire and analysed using multi-process theory. It was thought that the I-LED phenomenon would be reported by this cohort and be prevalent based on the routine nature of human error effects. This study found the cohort had experienced I-LED events during normal everyday aircraft maintenance operations, which appeared prevalent and made maintenance safer. Latent detections occurred more often At Work than Not at Work; thus, I-LED was found to be most successful when post-task schema housekeeping takes place in the same physical environment to that which the error occurred. This is new knowledge in the field of error detection. The concept of schema housekeeping also appears to be a new concept that contributes to schema theory, for which it was also argued that schema housekeeping is largely autonomous and persistent. Error events in the workplace were also detected away from the work environment; thus, this extends distributed cognition thinking beyond the need to remain proximal to the error event or physically immersed in the same system context to where the error actually occurred. Further, the argument that post-task schema housekeeping may be indiscriminate provides a new explanation for occasions of false alarms, or conversely chance detections. Notably, the very existence of I-LED indicates important cues remain available for post-task schema housekeeping to detect past errors, for which timing data from the group sessions identified a golden time window of two hours in which most I-LED events occur. The strength and distribution of system cues were also found to be important for schema triggering; thus, 'task-related' cues dominated At Work due to the high concentration of work-related cues in the aircraft maintenance environment. The findings further authorised the theoretical framework as effective for observing the I-LED phenomenon, for which time, location and other systems cues influence I-LED. Using a systems approach combined with multi-process theory advances current safety thinking. The findings from Objective 2 also provided direction to start identifying practicable interventions in support of the next objective.

10.2.3 Objective 3

Identify practicable interventions that enhance I-LED events in safety critical contexts.

Further data were collected from a new cohort of aircraft engineers during a diary study and analysed using multi-process theory linked to the Perceptual Cycle Model (PCM: Neisser, 1976). The intention was to advance knowledge of the nature and extent of I-LED events from a system perspective to identify practicable I-LED interventions. Additionally, the Cognitive Failures Questionnaire (CFQ: Broadbent et al., 1982) was administered. This simply confirmed that the sample of aircraft engineers exhibited normal cognitive safety behaviours associated with skilled workers; thus, the findings were likely to be transferrable to other populations of skilled workers. The study found I-LED events appear to mostly occur upon the deliberate review of past tasks within a golden time window of two hours of the error event occurring. Notably, this was during periods of unfocussed attention and whilst working alone in the same environment to that which the error occurred. Several sociotechnical factors associated with I-LED were studied and new practicable interventions identified. I-LED interventions were argued to be especially effective for simple everyday habitual tasks carried out alone where perhaps individual performance variability or human error effects are most likely to pass undetected if there are deficiencies in the safety system. Regaining Situational Awareness (SA) within the perceptual cycle through deliberate engagement with system cues supports the detection of past errors. Particularly, cues involving physical objects such as equipment or written words. I-LED interventions have been shown to help trigger the recall of past errors by deliberately engaging with system cues across the entire STS during the full range of normal behaviours, and therefore aids sufficient SA regain to detect past errors during the perceptual cycle.

A further review of literature found I-LED supports Safety II (Hollnagel, 2014) events by contributing to individual safety behaviour. Any I-LED intervention should be integrated within an organisation's safety system to maximise its benefit as an additional control, which also contributes to Safety I strategy. This finding supports Objective 5, which is discussed later. I-LED also contributes to knowledge of how to achieve greater safety resilience, for which it was recognised that the interventions identified in the diary study needed to be tested during normal operations in the workplace to explore their true effectiveness across everyday maintenance activities. This requirement formed the basis for the next study under Objective 4.

10.2.4 Objective 4

Understand the effectiveness of I-LED interventions in the workplace.

Based upon the findings from the study conducted in Chapter 6, several I–LED interventions were designed and tested for their effectiveness using a further cohort of naval aircraft engineers observed in their natural working environment (grouped operatives and supervisors). For the study presented in Chapter 7, it was hypothesised that the new interventions would be effective at triggering the recall of past errors within the time window of two hours. A Stop, Look and Listen (SLL) intervention was found to be most effective at triggering recall (out of four tested). It is thought the SLL intervention maximises an individual's engagement with system cues, especially visual cues. The other three interventions were ineffective for supervisors. The SLL intervention offers a new practicable safety control, as it is a flexible technique that the aircraft engineer can apply independently during normal maintenance activity. The tested I-LED interventions were argued to improve overall safety performance within an organisation, although it is recognised there are likely to be many more potential interventions than the four tested. This may be especially true for routine everyday aircraft maintenance operations where tasks are carried out alone or tasks are perceived to be low risk. I-LED research offers a further step-change in safety thinking by facilitating Safety II events through the application of I-LED interventions post-task completion. It was argued successful I-LED limits occasions for adverse outcomes to occur, despite the presence of existing safety controls designed to avoid or detect error effects. It is believed I-LED interventions enhance safety behaviours (or heroic recoveries) within safety critical sociotechnical contexts, as the interventions help guide users to engage with system cues to achieve the timely detection of latent error conditions. However, this new addition to established safety systems needs to be integrated within an organisation's overall safety strategy to enhance resilience, which is discussed under Objective 5.

10.2.5 Objective 5

Assess the benefit of integrating I-LED interventions with organisational safety strategies to enhance resilience.

To address the last objective, theoretical perspectives on optimising safety resilience were considered in Chapter 8 before reviewing the benefits of integrating I-LED interventions within a safety system in Chapter 9. It has been argued that achieving resilience through a Total Safety Management (TSM) approach to improving system safety reflects a progressive strategy that optimises sociotechnical controls at the organisational level with the management and delivery of error detection skills through individual safety behaviours in the workplace. A further literature review found few models describing an agreed format for managing organisational resilience based on total safety thinking. Thus, a hierarchical relationship described by a new TSM model was derived from various safety theories, including Human Factors Integration (HFI) and the 5M's concept (Harris & Harris, 2004). Theory highlighted the dichotomy between 'what should be done' and 'what can be done'. The latter could be expanded to highlight a further dichotomous relationship between safety 'as designed' by the organisation and 'as done' behaviour in the workplace (Hollnagel, 2014). It was argued the relationship between the elements shown in the model needs to be optimised to achieve total safety, although it is recognised that the simplified model belies the multiple interacting subsets and complex communication paths within a network of systems in systems that reflects the living and ever changing nature of an STS. Against the new TSM model, it was also argued that resilience is dependent on the system's ability to promote safe behaviours during interactions between humans and operating environments, which should be predicated on the presence of competent operators in the workplace. Reference to competence or expert operators was found throughout reviews of literature, but no definitive model was found that defined competence. Therefore, a new operator competence model was also constructed to help understand human-centric factors that have the potential to impact safety performance. I-LED success is believed to be dependent upon competence, especially non-technical skills (Flin et al., 2008), and therefore forms an essential component of the TSM model. The Safety II benefit comes from effective I-LED interventions that are used as an additional safety control at the local level by competent operators who are qualified, experienced and current for the task in hand, risk confident and able to maintain required performance despite being routinely surrounded by Performance Shaping Factors (PSFs) (Kirwan, 1998). To counter concerns over the term human error, it was highlighted that operator competence is dependent upon system controls. This was argued to be congruent with systems thinking and a total safety strategy, as the elements comprising the TSM require system controls (as designed) that are enacted locally (as done).

Arguably, the overall benefit of I-LED interventions comes from improved operator competence that underpins I-LED success through locally enacted examples of safe behaviours. This benefit is believed to support safety resilience provided I-LED interventions are integrated within an organisation's overall safety system and founded in demonstrable theory to form part of an enduring safety strategy. Benefits also include reduced injuries or death, avoiding equipment damage and economic gains as well as intangible socio-political effects and non-technical attributes such as improved operator empowerment and job satisfaction. However, Chapter 9 reviewed literature on the Cost versus Benefit Analysis (CBA) and Return on Investment (ROI) calculations for introducing new safety controls, which revealed detailed financial calculations were challenging due to often intangible cost data.

I-LED interventions tested in Chapter 7 do not require significant financial investment to integrate within the safety system as all were simple methods of focussing attention on system cues and, therefore, did not require significant investment in training, management or infrastructure. Based upon example occurrences reported in the Ministry of Defence (MoD) Air Safety Information Management System (ASIMS), representative costs for the SLL intervention were calculated using Health and Safety Executive (HSE) valuations for death and injuries. This showed that the significant CBA and ROI for high- and medium-event severities with low-severity events returning some financial benefits, whilst the calculation for negligible events showed a negative return. It has been argued, though, that I-LED interventions are likely to benefit the recovery from all levels of maintenance error events, regardless of severity, as organisational accidents are rarely the result of a single system failure. This is especially true for safe or ultra-safe organisations where high-risk hazards (reported as a high-event severity in ASIMS) are already subject to multiple system controls.

10.3 EVALUATION OF THE APPROACH TO RESEARCH

10.3.1 SAFETY RESEARCH STRATEGY

The observed population existed within the context of UK naval aircraft maintenance, which comprised 1700 operatives and supervisors, which comprised males and females of broad ethnicity, aged 18–50 years. In the study of human performance variability, only the context changes, which was discussed in Chapter 5. Using the Generic Error Modelling System (GEMS) categorisation of error (Reason, 1990), data from the study presented in Chapter 5 showed the category and ratio of errors reported by naval aircraft engineers to be broadly representative of wider aircraft maintenance error studies, and the application of schema theory has been shown in previous research to be influential and effective for military applications, including aviation. This provided confidence that the observed population was representative, and the findings from the current research should be transferrable to other populations of skilled operators.

Having identified the population, the challenge was to decide how to observe I-LED events in the population, either through experiment in a controlled laboratory setting and/or natural workplace environment. It was argued in Chapter 4 that understanding error effects comes from observing safety behaviours during normal operations where real-world studies produce the necessary insight on situational error from a systems perspective. Here, people create safety in the real world under resource and performance pressures at all levels within the STS, which is a view supporting both systemic and individual error contexts. Deliberately, this positioned current research within the realm of Human Factors and Ergonomics (HFE) (Carayon, 2006) research rather than cognitive psychology, as it is the influence of the external environment on safety behaviour that offers the greatest safety value. Ecological experiment also brings benefit with observational studies, since schema are internal representations of the world, for which measurement can only come from observed behaviour. It was recognised from the outset that research outside of a laboratory setting meant strict experimental control was not possible, although the advantage of naturalistic studies is that it avoids the bias of artificially controlled experiments that can erode the ecology of findings. Studies have also shown the nature of the operating context is not easily replicated in simulated experiments due to the complexity of sociotechnical interactions. However, the ecological benefits are recognised widely to outweigh any such concerns to deliver a meaningful contribution to safety knowledge.

A challenge to studying real-world events is selecting research methodologies that facilitate effective data gathering from the workplace. To help ensure quality data were captured, and to remain flexible to emergent findings, a staged approach using a mixture of methodologies was applied to a series of linked studies. This flexible approach facilitated the application of various research paradigms and data collection instruments to observe the target population over a protracted period and safeguarded against any individual study that failed. Emergent findings from each study were used to guide research, which also required continued engagement with current literature for the iterative development of hypothesis and theories. As I-LED events appear to be an under-researched safety field, there was little established theory to guide the research, and thus, the linked studies proved to be an appropriate method to develop thinking as findings emerged.

10.3.2 LIMITATIONS

The current safety research limited observations to naval aircraft engineers. However, literature reviews conducted in Chapters 2 and 6 showed similar human error types and rates to that seen in other safety critical organisations, whilst the CFQ scores recorded in Chapter 6 were representative of highly skilled operators, such as aircraft engineers. This suggests that the findings from this study should be generally applicable to other military and civilian organisations.

Very few I-LED examples were available from safety reports held in ASIMS. This was not unexpected, since an individual may not see the need to report an error that has been detected and recovered successfully. If applying Bird's (1969) theoretical safety triangle, there are likely to be many thousands of unreported safety-related events that occur in large organisations and thus many missed opportunities to analyse real world data from latent error events. The absence of safety reports also limited costs-versus-benefits calculations for the I-LED interventions introduced in Chapter 9.

Analysis of data from the self-report diaries described in Chapter 6 was limited to 51 usable I-LED events. Feedback from squadron engineering management highlighted that participant workload was extremely high, so they were not always able to make diary entries, whilst several engineers were re-employed away from the squadron, off sick or on short-notice courses. The nature of the operating environment also limited the time allowed to conduct the study and the number of questions that could be asked. These factors reiterated the challenge of observing normal operations in the workplace, but the ecological data that were gained provided the necessary insight into previously unaccounted safety behaviours. Typical limitations with collecting data were also experienced when testing I-LED interventions during the study described in Chapter 7. The shortfall in data collected from the observers limited the ability to analyse past error events detected via general Latent Error Searching (LES) as opposed to a specific cue contained in one of the booklets.

10.4 FUTURE WORK

The safety research presented in this book has focussed on individual safety behaviours since Chapter 5 found I-LED most likely to occur when alone. It is anticipated the I-LED phenomenon extends to teams, although it is not known what the impact on PCM performance would be as the physical interaction between operators may distract individual schema housekeeping. Thus, the benefit of I-LED in teams warrants study.

The low returns from the self-report diaries conducted over two months showed future research would benefit from study of a larger sample over a longer period in a cohort that is able to commit fully to making diary entries. Since the studies centred on highly skilled aircraft engineers, samples of non-aircraft or unskilled workers should also be considered. It was beyond the scope of the current safety research to observe all the potential sociotechnical factors that might contribute to successful I-LED events involving aircraft maintenance, both At Work and Not at Work. Thus, further research is needed to mature the exploratory nature of the current study to understand more about how cognition associated with the perceptual cycle is distributed across networks of sociotechnical contexts, in which I-LED events were seen to occur. It would also be advantageous to conduct I-LED research in other workplace contexts where further examples of I-LED interventions are likely to be identified. PCM performance can be influenced by the level of focal processing associated with an ongoing task; thus, the number of separate tasks completed during maintenance activities, strength of association between tasks and the 'difficulty' of each task may be influential in latent error detection and should also be researched further.

The PM element of multi-process theory highlights a preparation phase (encoding) when forming intention. Literature indicated that whether a PM task on the 'to-do' list is recalled correctly might be influenced by the quality of the preparation phase during encoding of the task. In their discussion on reflexive-associative theory relating to spontaneous recall of a PM, Einstein and McDaniel (2005) argued that processing of external cues to recall intentions is dependent on the extent the operator encoded the task: through mental task rehearsal, importance placed on the tasks and/or whether the task was to occur during a sequence of other tasks. Similarly, Kvavilashvili and Mandler (2004) argued association priming improved memory recall; thus, the quality of task encoding may also improve retrieval of information about past tasks when recall occurs. Specifically, it has been argued that operators normalise system hazards to such an extent that they are no longer cognitively prepared to detect and recover from error, and are therefore desensitised to risk (Reason, 1990). Regaining SA, and therefore facilitating I-LED success, might be improved through mental rehearsal (Annett, 2006; Flin et al., 2008), and thus, research is needed to explore whether I-LED performance is improved through task rehearsal. Similarly, there was a concern in Chapter 7 when testing I-LED interventions that operators would become desensitised to the intervention due to habit intrusion if used repeatedly over time. Further research is needed to determine whether habit intrusion through the routine everyday application of I-LED interventions counteracts its benefits over a protracted period of time. Conversely, Hutchins (1995) observed that every occasion of successful error detection provided the opportunity to develop individual error detection skills through schema development. Thus, the study of I-LED interventions over time should also determine if overall error detection improves either proximal or latent error detections.

It was argued error suppression strategies might generate too many safety processes that desensitise operators to important system cues, and it is likely to be difficult to control all safety situations with a process that may not reliably control all hazards impacting normative safety behaviours. Well-chosen I-LED interventions in lieu of overly prescriptive safety processes may improve overall safety performance and is therefore a new safety approach that would also benefit from further study.

10.5   CONCLUDING REMARKS

The purpose of the safety research presented in this book has been to explore the nature and extent of I-LED in aircraft maintenance and its contribution to making the safety system safer through enhanced resilience. Multi-process theory combined with systems thinking provided a theoretical framework upon which to observe the I-LED experiences of cohorts of naval aircraft engineers during normal everyday maintenance operations. The safety research was structured around five objectives, which first confirmed the presence of the I-LED phenomenon. Further findings showed time, location and other system cues facilitate I-LED events, for which the deliberate review of past activity within a time window of two hours of the error occurring and whilst remaining in the same sociotechnical environment to that which the error occurred appears most effective. The detection of work-related latent error conditions also occurred when in non-work environments such as at home or driving a car, which indicated distributed cognition extends across multiple sociotechnical networks. The nature of I-LED was also found common in simple everyday habitual tasks carried out alone where perhaps individual performance variability is most likely to pass unchecked, which gives rise to the potential for latent error conditions to manifest in the safety system. Testing of several practicable I-LED interventions, designed to focus operator attention on system cues such as objects or written words, showed a stop, look and listen intervention to be most effective at detecting past errors that lie hidden in the workplace.

Organisational accidents are rarely the result of a single system failure, and thus, safer systems require effective to controls to mitigate for the myriad of hazards that can lead to harm to people, damage equipment or reduce output. The I-LED interventions offer an additional safety control to help counter the potential risk escalation from undetected latent error conditions. A review of CBA and ROI literature showed that it is problematical to calculate the financial benefit of safety controls such as I-LED interventions due to a lack of tangible cost data, although the application of HSE valuations for deaths and injuries to typical maintenance error events reported in ASIMS allowed representative costs to be calculated. Whilst difficult to cost in financial terms, I-LED interventions tackle past errors that lie hidden and propagate through the entire STS, with the potential to network with other effects to create a causal path to harm or other undesirable outcome. Any safety control should be founded in theory and form part of an enduring long-term safety strategy to engender safe behaviours in the workplace to help avoid or reduce the risk of harm. The overall benefit of I-LED as an additional safety control offers physical benefits in terms of reduced injuries or death and equipment damage but also economics gains as well as socio-political effects and non-technical attributes such as improved operator situational awareness, stress and fatigue management, and job satisfaction. Analysis of UK military safety data also showed safety occurrences where the perceived severity was high or medium. It was argued that use of an I-LED intervention might have prevented the occurrence, and thus, they are thought to offer significant ROI to safety-critical organisations aiming to maximise the heroic abilities of its operators through enhanced safety competence. It has been argued the introduction of I-LED interventions as a long-term safety strategy does not attract significant financial costs or other resourcing implications within an organisation. I-LED interventions should also be integrated within the overall safety system to improve resilience, for which a new TSM model has been offered that describes a hierarchical relationship for total safety, predicated on the presence of competent operators with the technical and non-technical skills needed to mitigate for human error effects.

This book has explored improvements to workplace safety through the application of systems thinking and multi-process theory. It has been argued that I-LED interventions should make aircraft maintenance safer, especially if integrated within the safety system using a TSM approach. The population of aircraft engineers observed in the various studies has been shown to be typical of skilled operators, and therefore, the findings should translate readily to other populations of skilled operators where the safety dependent organisation seeks to enhance system safety resilience through improved operator competence in the workplace. The total effort of integrating new I-LED interventions with competent operators should yield a safer and more effective working environment, which offers a demonstrably return on investment.

Appendix A: Typical Maintenance Environments

Typical flight deck where maintenance is carried out. (Picture Crown Copyright ©.)

Maintenance hangar at sea. (Picture Crown Copyright ©.)

Maintenance hanger at a home base. (Picture Crown Copyright ©.)

Maintenance on flight deck at sea. (Picture Crown Copyright ©.)

Hooking load to helicopter at sea. (Picture Crown Copyright ©.)

Maintenance on flight deck at sea in extreme weather. (Picture Crown Copyright ©.)

Temporary maintenance hangar deployed. (Picture Crown Copyright ©.)

Replacing an aircraft engine at sea. (Picture Crown Copyright ©.)

AMCO at a home base. (Picture Crown Copyright ©.)

Deployed AMCO in field at night. (Picture Crown Copyright ©.)

Typical squadron crew room. (Picture Crown Copyright ©.)

Engine change in Norway. (Picture Crown Copyright ©.)

Appendix B: Participant Information Sheet (Chapter 5)

GROUP SESSIONS – PARTICIPANT INFORMATION SHEET

Study Title: Self-Detection and Recovery from Human Error

The following Participant Information Sheet has been prepared in accordance with Ministry of Defence (MoD) Research Ethics Committee (MoDREC) guidelines. Please read the following information carefully before agreeing to take part in this research.

What is the research about?

1. PhD research has been sponsored by the Royal Navy to explore how human error is self-detected and recovered across naval aviation. Research will consider naval air engineers initially and then branch out to other areas of the Fleet Air Arm.

Why have I been chosen?

2. As part of this research, you have been selected randomly from the squadron manpower list to participate in a focus group that will discuss your views on how everyday human error is self-detected and recovered amongst naval air engineers.

What will happen to me if I take part?

3. Each group session will consist of four air engineers from your squadron. After a brief from the researcher, you will be asked to complete an attendance register and consent form. The register will capture the following information, which is needed to help analyse the collected data and will be kept separate from your signed consent form:
   • Rank
   • Trade
   • Age
   • Sex
   • Time in Service

4. The group session will take no longer than 30 minutes. It will involve a discussion on everyday human error at work. You will also be asked to write down your answers to questions on an example of human error that happened to you. For this, **you will need to come to the group with an example of an everyday error that happened to you at work**. The example needs to be an error that you did not notice at the time it actually occurred but that you later detected after the job/task was completed. The error must

be one that you self-detected and not one which was discovered by a third party (i.e., a supervisor) or from following a set process. NB: For the study, error is defined simply as 'not doing what the situation required'.

Are there any benefits in my taking part?

5. Your participation will allow you to make an important contribution to the understanding of human error in our Federal aviation administration (FAA) so that practical interventions can be designed to help improve the self-detection of error. Updates will be published in the RN Flight Safety magazine, and your contribution will be combined with other initiatives/ research on error that you may have heard of.

Are there any risks involved?

6. This research has been assessed as low risk using MoD Research Ethics Committee guidelines, and Southampton University has provided general ethical approval. Your involvement does not pose any physical or psychological danger and will not cause any impact on your career. As a reminder, the Royal Navy operates a Just Culture that applies equally to this research when providing essential safety-related information. **Please note that you are not to discuss any error that resulted in mandatory occurrence reporting for injury to personnel, aircraft damage or where disciplinary action was taken.**

Will my participation be confidential?

7. Information obtained from your involvement will be treated in strict confidence. Your personal details and specific contribution will not be available to Command or any other party. To preserve anonymity, your name will not be recorded with data to be analysed, and all identifying information relating to organisations, squadrons, specific locations and aircraft will not be published. The researcher may need to take notes, which you are free to view. Additionally, your consent form will be held by your AEO and not by the researcher.

What happens if I choose not to participate?

8. You do not have to participate in this research if you do not want, for which you do not need to give a reason why.

Where can I get more information?

9. Your AEO has been briefed on the purpose of this research and the scope of the focus group. The researcher can be contacted as follows:

Appendix C: Group Sessions – Themes Derived from Question 9 (Chapter 5)

Narrative Response (n = 48)	Themes (n = 48)
Reviewing task in head/doubt	Reflection/review
Just thinking/disbelief in work	Self-doubt/suspicion
Self-doubt/unsure	Self-doubt/suspicion
Came to mind watching TV	Came to mind
Occurred something not correct	Came to mind
Questioned in mind after work	Reflection/review
Just came to my head	Came to mind
Feeling something amiss after secure	Self-doubt/suspicion
Came to mind at lunch	Came to mind
Recalled when completing paperwork	Task-related cue
Noticed associated tool then realised	Task-related cue
When discussing work with colleague	Discussing work
Triggered using similar equipment	Task-related cue
Noticed HUMS computer and realised	Task-related cue
Realised when reviewing work (in head)	Reflection/review
Niggling feeling	Self-doubt/suspicion
Suspicious signing paperwork	Task-related cue
Realisation just hit me	Came to mind
Niggling feeling not replaced all plugs	Self-doubt/suspicion
Uneasy feeling after work	Self-doubt/suspicion
During handover discussion occurred to me	Discussing work
Triggered by similar task on other aircraft	Task-related cue
Knew something wrong so checked	Self-doubt/suspicion
Came to mind seeing hydraulic rig	Task-related cue
Triggered by incident on another aircraft	Task-related cue
Came to mind when partner asked about day. Initial suspicion only	Self-doubt/suspicion
Came to mind	Came to mind
Thinking about day	Reflection/review
Niggling feeling something not right	Self-doubt/suspicion
Reviewed tasks in head after work	Reflection/review
Felt uneasy when watching TV	Self-doubt/suspicion
Remembered when doing next stage	Task-related cue
Uneasy feeling clearing away	Self-doubt/suspicion
Was discussing job at stand easy	Discussing work

(Continued)

137

Narrative Response (n = 48)	Themes (n = 48)
Remembered walking to AMCO to sign	Came to mind
Flash mental image of ac jacks that led to thinking of adapters	Came to mind
Remembered looking out window	Came to mind
Visualised job and remembered error	Reflection/review
Mental review during pause in operations	Reflection/review
Thinking about job before sleeping	Reflection/review
Remembered check by chance	Came to mind
Reflecting on work, came to mind	Reflection/review
Thinking about task then unsure checked/became suspicious	Reflection/review
Saw Sea King on news winching and came to mind high lines not packed	Task-related cue
Entering numbers in 700C made me question if ac number correct	Task-related cue
When completing same task on other ac caused to question	Task-related cue
Nagging feeling forgotten something then came to mind	Self-doubt/suspicion
Came to mind when catching up on paperwork	Came to mind

Appendix D: Group Sessions – Themes Derived from Question 13 (Chapter 5)

Narrative Response (n = 42)	Themes (n = 55)
Trying to minimise interruptions. Potentially having a break between tasks to allow reflection on tasks completed.	Avoid interruptions/distractions Rest/break needed between tasks
Stop doubting yourself. Check things more than once.	Check work
Q&A before finishing work by supervisor to detect error prior to leaving.	Check work
Having a checklist for each task to run through after completion.	Use checklist/process
I am not sure – it is human nature to make errors. Processes are in place to minimise the risk of this.	Use checklist/process
To quickly double check the process carried out.	Check work
Avoid feeling rushed at work, although I feel this is mostly my perception and not the intention.	Avoid rushing
More hands-on training instead of so much classroom work.	Training must be fit for purpose
Double check work before signing for it. Taking your time with your tasks.	Check work Avoid rushing
By constant practice. However, due to broadening of responsibility by covering a wide range of different systems, skill fades and a lack of depth of knowledge will mean that self-detection suffers.	Experience/skill needed
Run a task through your head on completion.	Check work
By more careful self-assessment.	Check work
Greater depth of training for Phase 2Bs (air engineer training).	Training must be fit for purpose
Don't be complacent or overconfident and be conscious of how others perceive your work output.	Error awareness
By reading through instructions as you work, rather than before a long task then afterwards, just to be sure nothing was missed.	Use checklist/process
A list of common errors to check for on the work cards.	Use checklist/process
Encouraging people to check their work until they are absolutely certain that it has been carried out correctly.	Check work
I believe self-detection is a personality thing. It is how fastidious a person is at that particular moment. I am unsure how it could be improved.	Influence of personality
Possibly some kind of system to enable each person on the job to go through all the jobs for the shift.	Check work
Possibly having a short break (5 or 10 minutes) between jobs/tasks to give time to think and not get clustered/confused in your head if you are carrying out the same task one after the other repeatedly.	Rest/break needed between tasks

(Continued)

Narrative Response (n = 42)	Themes (n = 55)
Maybe notifications or bold parts in Topics to jog your memory of what should be done/should have been done.	Use warnings
We could be made more aware that we are prone to errors.	Error awareness
I think experience plays a big part of self-detection and the ability to openly admit or discuss potential errors without the fear of reprisals or punishment for making an error or openly admitting it.	Experience/skill needed Just Culture
Spending a period of time at the end of each working day reflecting on how day went.	Check work
If a task seems repetitive, perhaps take extra care as it can be the simple tasks that cause more issues.	Check work
Going over the task before, during and after helps.	Check work
Details on paperwork should be thorough, and emphasis on paperwork should be a training factor on all career courses.	Processes must be fit for purpose Training must be fit for purpose
Employ better engineers or improve perks. People don't care about an employer that doesn't care genuinely.	Experience/skill needed Reward/incentives Employer responsibility
By removing some of the pressure to make slots and rushing jobs allowing for time to check twice and allowing for personnel to take breaks (quick 5mins breather) without being made to feel lazy.	Avoid task pressure Check work Rest/break needed between tasks
Not really sure as everyone is different, and some remember and others do not.	Influence of human performance
By making sure you understand fully what job/task you're undertaking and by conducting it straight away.	Experience/skill needed Avoid delays
One-man one-job. Avoid overloading tasks/people. Culture change on Watch to work a set time, not work towards early 'chop'.	Avoid tasks in parallel Avoid task pressure Avoid interruptions/distractions
Having time post task to reflect on it.	Rest/break needed between tasks
Constantly check periodically what you have done by referring to publications. Have tick sheets to confirm task had been complete.	Use checklist/process
Try to avoid distractions during jobs. One job at a time until its completion.	Avoid tasks in parallel Avoid interruptions/distractions
Responsibility and empowerment should be pushed/delegated to the lowest level possible in order to give people ownership of the task. If they feel they own the task, they will be more likely to do it properly and reflect on their own work.	Individual responsibility
Less pressure to achieve tasks in busy periods.	Avoid task pressure
Reduce perceived pressure supervisors think they are under.	Avoid task pressure
Know that you are capable of error and regularly self-check even minor tasks to eliminate grey outs.	Error awareness Check work
Allowing more time to complete tasks.	Avoid task pressure
Self-discipline and don't get sidetracked.	Avoid interruptions/distractions
Very difficult. If the job was important, you naturally check and check again to ensure it is completed correctly. Emphasising the importance of tasks may help; however, if we over emphasise everything, you would lose the effect (i.e., making everything important).	Emphasise task importance Check work

Appendix E: Cognitive Failures Questionnaire (Chapter 6)

DIARY STUDY – COGNITIVE QUESTIONNAIRE

Participant Number: _____

The following questions are about minor errors that everyone makes from time to time, but some of which happen more often than others. Please circle one of the numbers (0–4) for each question, which should cover the last 6 months (Answer questions 1–25):

		Very Often	Quite Often	Occasionally	Very Rarely	Never
1.	Do you read something and find you haven't been thinking about it so must read it again?	4	3	2	1	0
2.	Do you find you forget why you went from one part of the house to the other?	4	3	2	1	0
3.	Do you fail to notice signposts on the road?	4	3	2	1	0
4.	Do you find you confuse right and left when giving directions?	4	3	2	1	0
5.	Do you bump into people?	4	3	2	1	0
6.	Do you find you forget whether you've turned off a light or a fire or locked the door?	4	3	2	1	0
7.	Do you fail to listen to people's names when you are meeting them?	4	3	2	1	0
8.	Do you say something and realize afterwards that it might be taken as insulting?	4	3	2	1	0
9.	Do you fail to hear people speaking to you when you are doing something else?	4	3	2	1	0
10.	Do you lose your temper and regret it?	4	3	2	1	0
11.	Do you leave important letters unanswered for days?	4	3	2	1	0
12.	Do you find you forget which way to turn on a road you know well but rarely use?	4	3	2	1	0
13.	Do you fail to see what you want in a supermarket (although it's there)?	4	3	2	1	0

(Continued)

	Very Often	Quite Often	Occasionally	Very Rarely	Never
14. Do you find yourself suddenly wondering whether you've used a word correctly?	4	3	2	1	0
15. Do you have trouble making up your mind?	4	3	2	1	0
16. Do you find you forget appointments?	4	3	2	1	0
17. Do you forget where you put something like a newspaper or a book?	4	3	2	1	0
18. Do you find you accidentally throw away the thing you want and keep what you meant to throw away – such as throwing away the contents of a package you wanted but keeping the packaging?	4	3	2	1	0
19. Do you daydream when you ought to be listening to something?	4	3	2	1	0
20. Do you find you forget people's names?	4	3	2	1	0
21. Do you start doing one thing at home and get distracted into doing something else (unintentionally)?	4	3	2	1	0
22. Do you find you can't quite remember something although it's 'on the tip of your tongue'?	4	3	2	1	0
23. Do you find you forget what you went to the shops to buy?	4	3	2	1	0
24. Do you accidentally drop things?	4	3	2	1	0
25. Do you ever find you can't think of anything to say?	4	3	2	1	0

Appendix F: Self-Report Diary (Chapter 6)

POST TASK ERROR DETECTION EVENT: 1	DATE: _____

Q1. Please give a brief description of the error event:

Q2. At what time did the error event occur? _____ (Exact time or within nearest 30mins)

Q3. What type of task was it? Complex ☐ Simple ☐ NK ☐

Q4. What was the cue to do this task? Event ☐ Time ☐ Both ☐

Q5. What was the error type? Slip ☐ Lapse ☐ Mistake ☐ NK ☐

Q6. Where were you when the <u>error occurred</u>?

- AMCO ☐
- Hangar ☐
- Line ☐
- Maintenance office ☐
- Issue centre ☐
- Storeroom ☐
- In aircraft ☐
- Workshop ☐
- Flight deck ☐
- Other (please specify) ☐

Q7. At what time did you recall the error (post task completion)? _____ (Exact time or within nearest 30mins)

Q8. Where were you when you <u>recalled the error</u>?

- AMCO ☐
- Hangar ☐
- Line ☐
- Maintenance office ☐
- Crew room ☐
- Issue centre ☐
- Storeroom ☐
- In aircraft ☐
- Workshop ☐
- Flight deck ☐
- Home/Mess ☐
- In bed ☐
- In a vehicle ☐
- Gym ☐
- Other (please specify) ☐

Q9. What were you doing when you recalled the error?

- Planning/preparing maintenance activity ☐
- Conducting similar maintenance activity ☐
- Conducting dissimilar maintenance activity ☐
- Walking ☐
- Driving a vehicle ☐
- Exercising (i.e., cycling, running, etc) ☐
- Showering ☐
- Eating ☐
- Socialising (i.e., in a pub) ☐
- Discussing work (not formal handover/brief) ☐
- Daydreaming ☐
- Resting ☐
- Entertainment (i.e., reading, TV, internet, etc) ☐
- Sleeping ☐
- Other (please specify) ☐

Q10. Did you intentionally review your past tasks/activities? Yes ☐(Go to Q11) No ☐(Go to Q12)

Q11. Was this part of your personal routine? Yes ☐ No ☐

Q12. On checking your work, was the error: Real ☐ False Alarm ☐

Q13. Did anything in your immediate location appear to trigger the error recall?

- Sound ☐
- Equipment ☐
- Document ☐
- Smell ☐
- Taste ☐
- General vista ☐
- Other ☐

Please describe:

Q14. What were you thinking about at the time of the error recall? (Answer either 14a or 14b)

Q14a. Work-related thoughts:	Q14b. Non work-related thoughts:
• Past task / event ☐ • Task in-hand ☐ • Future task / event ☐ Please describe:	• Past activity / event ☐ • The 'moment' ☐ • Future activity / event ☐ Please describe:

Q15. Were you alone when the error was recalled?   Yes ☐   No ☐

	Strongly disagree	Disagree	Uncertain	Agree	Strongly agree
	5	4	3	2	1
Q16. The specific error was **very clear** to me.					
Q17. I was **very confident** that my past task was in error.					
Q18. The error recall occurred when I was **highly focused** on the activity at Q9.					

Please use for making notes or to add any additional information:

Appendix G: Participant Information Sheet (Chapter 6)

DIARY STUDY – PARTICIPANT INFORMATION SHEET

Study Title: Post-Task Latent Error Detection in UK Naval Aircraft Maintenance

The following Participant Information Sheet has been prepared in accordance with Ministry of Defence (MoD) Research Ethics Committee (MoDREC) guidelines. Please read the following information carefully before agreeing to take part in this research.

What is the purpose of the research?
Human error is inevitable and a daily occurrence and is the most significant factor in aircraft safety-related occurrences. This is well known in the Fleet Air Arm, for which we are very proactive in removing error (or reducing the likelihood) through the competency of our people and via careful adherence to rules and procedures. A task carried out in error inadvertently creates a latent error that can result in a future undesirable outcome if the error is not detected later. Detection of typical latent errors, post-task completion, has been observed amongst UK naval air engineers and is reported to be a result of some seemingly spontaneous recollection of past activity. This diary study is trying to understand the nature and extent of latent errors that are detected post-task completion so that interventions can be designed to help enhance their detection.

Who is doing this research?
The researcher is a serving Air Engineer Officer who is sponsored by the Royal Navy.

Why have I been invited to take part?
Research is looking at naval air engineers initially but will branch-out to other areas of the Fleet Air Arm. As part of this research, you have been selected randomly from the squadron manpower list to participate in an anonymous diary study that will record the self-detection of latent errors.

Do I have to take part?
You do not have to participate in this research if you do not want to.

What will I be asked to do?
You will be briefed on the purpose of the research and how your data will be used before starting the study. The researcher will give the brief. After the brief, you will be asked to complete a simple questionnaire to baseline the study, a participant

register and consent form. The register will capture the following information, which is needed to help analyse the collected data: rank, trade, age and sex.

The register will be kept separate from your signed consent form. You will be given an anonymous participant number to write on a diary booklet that you will be given. It should take no longer than two minutes to complete the questions in the diary for each occasion you experience a post-task latent error detection event. The study will take place in your normal work environment and will continue until you have recorded 5–10 latent error examples, or a maximum of two months has passed. There are no additional time or travel commitments, and no errors that result in injury to personnel or damage to equipment are to be recorded in the diary. **NB: Normal occurrence reporting procedures remain extant throughout**.

When you have completed 5–10 error examples, you will mail your diary direct to the researcher in the supplied pre-paid envelope.

What is the device or procedure that is being tested?

There are no devices or procedures under test other than those already available in your normal workplace environment.

What are the benefits of taking part?

The aim of this research is to help enhance the self-detection of latent errors that could otherwise pass undetected. Your participation will make an important contribution to the understanding of how latent errors are detected.

What are the possible disadvantages and risks of taking part?

This research has been assessed as low risk using the MoD Research Ethics Committee guidelines, and Southampton University has provided ethical approval. Your involvement does not pose any danger and will not cause any impact on your Service career. As a reminder, the Royal Navy operates a just culture that applies equally to this research when providing essential safety-related information.

Can I withdraw from the research and what will happen if I don't want to carry on?

You are free to withdraw from the diary study at any time and you are not required to give a reason.

Are there any expenses and payments that I will get?

The diary study is voluntary and thus no additional incentives are available.

Whom do I contact if I have any questions or a complaint?

The researcher can be contacted directly if you have any questions. In the case of a complaint or other concern, please contact your AEO in the first instance.

What happens if I suffer any harm?

This research has been assessed as low risk. You should not be exposed to any physical harm through this diary study and you are not to record any latent error examples that may cause you stress.

What will happen to any samples I give?
You are not required to provide samples.

Will my records be kept confidential?
All information provided will be treated in strict confidence. To preserve anonymity, your name will not be recorded with the collected data, and any identifying information relating to organisations, squadrons, specific locations and aircraft will not be published.

Who is organising and funding the research?
Research is organised by the researcher who is a serving AEO. The Royal Navy has funded this research.

Who has reviewed the study?
Southampton University Ethics Committee, Navy Command HQ and the RNFSC have reviewed this study.

Where can I get further information and contact details?
Your AEO has been briefed on the purpose of this research and scope of the diary study. The researcher can be contacted as follows:

Appendix H: Participant Information Sheet (Chapter 7)

INTERVENTION STUDY – PARTICIPANT INFORMATION SHEET (PIS)

Study Title: Individual Latent Error Detection in Naval Aircraft Maintenance

The following Participant Information Sheet (PIS) has been prepared in accordance with Ministry of Defence (MoD) Research Ethics Committee (MoDREC) guidelines. Please read the following information carefully before taking part in this research.

What is the purpose of the research?
Human error is inevitable and the most significant factor in aircraft-related safety occurrences. We all suffer errors, and this is a daily occurrence, for which a maintenance task that is inadvertently carried out in error creates a latent error that can result in a future safety issue if the error is not detected later. Air engineers have been shown to self-detect their latent errors through the recall of past activity. This happens <u>post-task completion</u> and can appear to be completely spontaneous. The self-detection of latent errors post-task completion is completely separate to detecting errors by following a process, mandatory checklist or via a supervisory check. This study is testing several intervention techniques designed to help promote the recall of latent errors, which may be present in completed maintenance activities.

Who is doing this research?
The researcher is a serving AEO who is sponsored by the Royal Navy through their elective studies programme with Southampton University.

Why have I been invited to take part?
The RNAESS is being used to test several intervention techniques using students attending career courses. This is so that the interventions can be tested in a controlled and safe environment prior to use in operating squadrons.

What will I be asked to do?
The study will take place at 760 and 764 Squadrons where training will be carried as normal. After completing a practical task, you may be asked to try an intervention technique. Your instructor will explain the intervention to you, which will be separate to the rules and procedures you have already been taught as part of your course. You must be alone when you try the technique to avoid distraction, after which you report back to your instructor who will ask you a series of short questions to gauge how useful you found the technique. Each technique is very simple, quick and will not impact any part of your course assessment. Your data will then be used to determine which technique is most effective and could be used in operating squadrons.

You will be briefed on the purpose of the research and how your data will be used. After the brief, you will be asked to complete a simple questionnaire to baseline this study against other studies that have analysed human error. Southampton University and MoD ethics committees require a consent form to be completed, which simply records that you agreed to participate in this study and so it will not be linked to your data as there is no requirement to do this. Your course officer will also complete a register of participants to record the course number, rate, sex and age of participants, which is needed to analyse collected data. As your data is anonymous, your course officer will issue you a participant number, which you need to include with your feedback to your instructor.

What is the device or procedure that is being tested?
There are no devices or procedures under test other than those already available in your normal workplace environment.

What are the benefits of taking part?
The aim of this study is to help enhance individual detection of latent errors that could otherwise pass undetected in routine everyday maintenance-related tasks, despite doing your best to avoid an error. Your feedback on the intervention techniques will help decide which one(s) should be used in live squadrons.

What are the possible disadvantages and risks of taking part?
This research has been assessed as low risk using the MoD Research Ethics Committee guidelines, and Southampton University has provided ethical approval. Your involvement does not pose any danger and will not cause any impact on your training or practical assessments.

Can I withdraw from the research, and what will happen if I don't want to carry on?
You do not have to participate in this research if you do not want to and you are free to withdraw from the study at any time.

Are there any expenses and payments that I will get?
The study is voluntary and within the scope of your career course; thus, no additional incentives are available.

Whom do I contact if I have any questions or a complaint?
The researcher can be contacted directly if you have any questions. In the case of a complaint or other concern, please contact your course officer in the first instance.

What happens if I suffer any harm?
This research has been assessed as low risk. You should not be exposed to any physical harm or stress through this study.

What will happen to any samples I give?
You are not required to provide samples.

Will my records be kept confidential?
All information provided will be treated in strict confidence. To preserve anonymity, your name will not be recorded with your data and any identifying information will be removed.

Who is organising and funding the research?
Research is organised by the researcher who is a serving AEO. The Royal Navy has funded this research under its elective studies programme.

Who has reviewed the study?
Southampton University Ethics Committee, RNAESS MoD and RNAESS CO have reviewed this study.

Where can I get further information and contact details?
Your course officer has been briefed on the purpose of this research and scope of the study. The researcher can be contacted as follows:

Appendix I: Example Picture Flash Card for an Operative (Chapter 7)

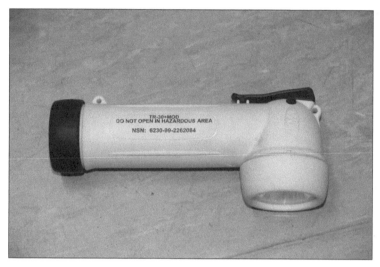

(Picture Crown Copyright ©.)

Appendix J: Example Picture Flash Card for a Supervisor (Chapter 7)

(Picture Crown Copyright ©.)

Appendix K: Example Word and Picture Flash Card for an Operative (Chapter 7)

Filler cap. (Picture Crown Copyright ©.)

Appendix L: Example Word and Picture Flash Card for a Supervisor (Chapter 7)

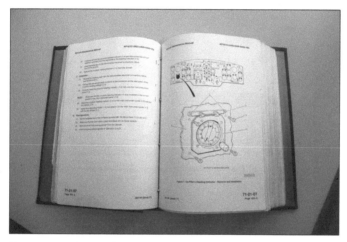

Maintenance checks. (Picture Crown Copyright ©.)

Appendix M: I-LED Study – Observer Form (Chapter 7)

SQUADRON:	COURSE NO:
PARTICIPANT NO:	DATE:
TASK:	NO. OF PREDICTED ERRORS:
INTERVENTION USED:	BOOKLET NO. (if applicable):

Please check the participant took their time with the intervention and were alone.

Q1. Number of participant errors observed by instructor:

Q2. Number of latent errors detected by the participant using the intervention:

Q3. For each detected latent error, please record the specific cue (contained within the intervention technique) that triggered the latent error recall. Specific cues are 'stop', 'look', 'listen', a particular word or picture.

Cue	Latent Error Detected	Cue	Latent Error Detected
E.g., Filler cap	Oil filler cap left off	3.	
1.		4.	
2.		5.	

Q4. Time lapse between task completion and intervention technique? _____ (Exact time or within nearest 5 mins)

Q5. Where did the participant try the intervention?

- AMCO ☐
- Hangar ☐
- Line ☐
- Maintenance office ☐
- Crew room ☐
- Issue centre ☐
- Storeroom ☐
- In aircraft ☐
- Workshop ☐

- Passageway ☐
- Classroom ☐
- Other (please specify) ☐

Q6. On checking their work, was the latent error:	Error 1: Real ☐ False Alarm ☐
	Error 2: Real ☐ False Alarm ☐
	Error 3: Real ☐ False Alarm ☐
	Error 4: Real ☐ False Alarm ☐
	Error 5: Real ☐ False Alarm ☐

Q7. Participant alone when using the intervention?   Yes ☐   No ☐

Q8. The intervention was effective: Yes ☐   No ☐

Please use for any additional information:

References

Adams, D. 2006. A layman's introduction to human factors in aircraft accident and incident investigation: ATSB Safety Information Paper-B2006/0094. Available from: https://www.atsb.gov.au/publications/2006/b20060094/

Aini, M.S. & Fakhru'l-Razi, A. 2013. Latent errors of socio-technical disasters: A Malaysian case study. *Safety Science* **51**, 284–292.

Air Accident Investigation Branch (AAIB). 1990. *BAC 1-11, G-BJRT.* Report No: 1/1992. Report on the accident to BAC One-Eleven, G-BJRT, over Didcot, Oxfordshire on 10 June 1990. Report No: 1/1992. Available from: https://www.gov.uk/aaib-reports/1-1992-bac-one-eleven-g-bjrt-10-june-1990.

Air Accident Investigation Branch (AAIB). 2000. Airbus A320-231, G-VCED: AAIB Bulletin No: 7/2000-EW/C2000/1/2. Available from: http://www.aaib.gov.uk/cms_resources.cfm?file=/dft_avsafety_pdf_501061.pdf

Air Accident Investigation Branch (AAIB). 2015. *Airbus A319-131, G-EUOE*: Aircraft Accident Report 1/2015 – Airbus A319-131, G-EUOE, 24 May 2013. Report No. 1/2015. Available from: https://www.gov.uk/aaib-reports/aircraft-accident-report-1-2015-airbus-a319-131-g-euoe-24-may-2013.

Allwood, C.M. 1984. Error detection processes in statistical problem solving. *Cognitive Science* **8**, 413–437.

Amalberti, R. 2001. The paradoxes of almost totally safe transportation systems. *Safety Science* **37**(2–3), 109–126.

Amalberti, R. 2013. *Navigating Safety: Necessary Compromises and Trade-Offs-Theory and Practice.* Heidelberg: Springer.

Amalberti, R. & Barriquault, C. 1999. Fondements et limites du retour d'expérience [Foundations and limits of feedback]. *Annales des Ponts et Chaussés* **91**, 67–75.

Amalberti, R. & Wioland, L. 1997. Human error in aviation. In *Aviation Safety*, H. Soekkha (ed.). Paper presented at the *International Aviation Safety Conference. Rotterdam Airport*, The Netherlands, 27–29 August 1997. 91–108. Utrecht: VSP.

Annett, J. 2006. *The Oxford Human Companion to the Mind.* Oxford: Oxford University Press, second edition.

Aviation Safety Information Management System (ASIMS). 2013a. Online database (restricted access). (Accessed 13 July 2013).

Aviation Safety Information Management System (ASIMS). 2013b. *User guide.* Available from: http://www.maa.mod.uk/linkedfiles/occurrence_reporting/20111005asims_user_guide_v42_finalu.pdf. (Accessed 30 October 2013).

Aviation Safety Information Management System (ASIMS). 2017. Online database (restricted access). (Accessed 17 January 2017).

Baddeley, A. 1997. *Human Memory: Theory and Practice.* Hove: Psychology Press, revised edition.

Baddeley, A. & Wilkins, A. 1984. Taking memory out of the laboratory. In *Human Memory: Theory and Practice*, A. Baddeley (ed.). 1997. Hove: Psychology Press, revised edition.

Bargh, J.A. & Chartrand, T.L. 1999. The unbearable automaticity of being. *American Psychologist* **54**, 462–479.

Bartlett, F.C. 1932. *Remembering: A Study of Experimental and Social Psychology.* Cambridge, MA: Cambridge University Press.

Bird, F.E. 1969. Practical loss control leadership. Institute Publishing (Division of International Loss Control Institute), Loganville, GA. In *Managing the Risks of Organizational Accidents*, J. Reason (ed.). 1997. Aldershot: Ashgate.

Blavier, A., Rouy, E., Nyssen, A.S. & De Keyser, V. 2005. Prospective issues for error detection. *Ergonomics* **48**(7), 758–781.

Brewer, G.E., Knight, J.B., Marsh, R.L. & Unsworth, N. 2010. Individual differences in event-based prospective memory: Evidence for multiple processes supporting cue detection. *Memory & Cognition* **38**(3), 304–311.

Bridger, R.S., Brasher, K., Dew, A., Sparshott, K. & Kilminster, S. 2010. Job strain related to cognitive failure in Naval personnel. *Ergonomics* **53**, 739–747.

Broadbent, D.E., Cooper, P.F., FitzGerald, P. & Parkes, K.R. 1982. The Cognitive Failures Questionnaire (CFQ) and its correlates. *British Journal Of Clinical Psychology* **21**(1), 1–16.

Campbell, R.D. & Bagshaw, M. 2002. *Human Performance and Limitations*. Abingdon: Blackwell, third edition.

Carayon, P. 2006. Human factors of complex sociotechnical systems. *Applied Ergonomics* **37**, 525–535.

Carayon, P., Hancock, P., Leveson, N., Noy, Y.I., Sznelwar, L. & van Hootegem, G. 2015. Advancing a sociotechnical systems approach to workplace safety: Developing the conceptual framework. *Ergonomics* **58**(4), 548–564.

Cassell, C. & Symon, G. 2004. *Essential Guide to Qualitative Methods in Organizational Research*. London: Sage.

Chatzimichailidou, M.M., Stanton, N.A. & Dokas, I.M. 2015. The concept of risk situation awareness provision: Towards a new approach for assessing the DSA about the threats and vulnerabilities of complex socio-technical systems. *Safety Science* **79**, 126–138.

Cheng, C.-N. & Hwang, S.-L. 2015. Applications of integrated human error identification techniques on the chemical cylinder change task. *Applied Ergonomics* **47**, 274–284.

Chiu, M.-C. & Hsieh, M.-C. 2016. Latent human error analysis and efficient improvement strategies by fuzzy TOPSIS in aviation maintenance tasks. *Applied Ergonomics* **54**, 136–147.

Civil Aviation Authority (CAA). 2009. Aircraft maintenance incident analysis. CAA Paper 2009/05. Civil Aviation Authority, UK. Norwich: TSO.

Cohen, G., Eysenck, M.W. & LeVoi, M.E. 1986. *Memory*. Milton Keynes: Open University Press.

Cooper, M.D. & Phillips, R.A. 1995. Killing two birds with one stone: Achieving quality via total safety management. *Leadership and Organisational Development Journal* **16**(8), 3–9.

Cornelissen, M., Salmon, P.M., Jenkins, D.P. & Lenné, M.G. 2013. A structured approach to the strategies analysis phase of cognitive work analysis. *Theoretical Issues In Ergonomics Science* **14**(6), 546–564.

Day, A.J., Brasher, A. & Bridger, R.S. 2012. Accident proneness revisited: The role of psychological stress and cognitive failure. *Accident Analysis And Prevention* **49**, 532–535.

Dekker, S.W.A. 2003. Illusions of explanation: A critical essay on error classification. *The International Journal of Aviation Psychology* **13**(2), 95–106.

Dekker, S.W.A. 2006. *The Field Guide to Understanding Human Error*. Aldershot: Ashgate, second edition.

Dekker, S.W.A. 2014. *The Field Guide to Understanding Human Error*. Aldershot: Ashgate, third edition.

Dhillion, B.S. 2009. *Human Reliability, Error and Human Factors in Engineering Maintenance*. Boca Raton, FL: CRC Press.

Dismukes, R.K. 2012. Prospective memory on workplace and everyday situations. *Current Directions in Psychology Science* **21**(4), 215–220.

Dockree, P.M. & Ellis, J.A. 2001. Forming and cancelling everyday intentions: Implications for prospective remembering. *Memory & Cognition* **29**(8), 1139–1145.

Doireau, P., Wioland, L. & Amalberti, R. 1997. Human error detection by outside observers: The case of aircraft piloting. *Le Travail Humain* **60**(2), 131–153.

Ebbinghaus, H. 1885:1964. *Memory: A Contribution to Experimental Psychology.* New York: Dover Publications.

Edwards, E. 1972. Man and machine: Systems for safety. In *Proceedings of British Airline Pilots Association Technical Symposium*, 21–36. London: British Airline Pilots Association.

Einstein, G.O. & McDaniel, M.A. 2005. Prospective memory: Multiple retrieval processes. *Current Directions in Psychological Science* **14**(6), 286–290.

Einstein, G.O., Smith, R.E., McDaniel, M.A. & Shaw, P. 1997. Aging and prospective memory: The influence of increased task demands at encoding and retrieval. *Psychology and Aging* **12**, 479–488.

Ellis, J. 1996. Prospective memory for the realization of delayed intentions: A conceptual framework for research. In *Prospective Memory: Theory and Applications*, M. Brandimonte et al. (eds.). Mahwah, NJ: Lawrence Erlbaum Associates, 1–23.

Emery, E.F. & Trist, E.L. 1960. Sociotechnical systems. In *Management Science*, C.W. Churchman & M. Verhulst (eds.). Oxford: Pergamon, second edition, 179–202.

Endsley, M.R. & Robertson, M.M. 2000. Situation awareness in aircraft maintenance teams. *International Journal Of Industrial Ergonomics* **26**, 301–325.

Fawcett, T. 2006. An introduction to ROC analysis. *Pattern Recognition Letters* **27**, 861–874.

Fedota, J. & Parasuraman, R. 2009. Neuroergonomics and human error. *Theoretical Issues in Ergonomics Science* **11**(5), 402–421.

Finomore, V., Matthews, G., Shaw, T. & Warm, J. 2009. Predicting vigilance: A fresh look at an old problem. *Ergonomics* **52**(7), 791–808.

Fitts, P.M. & Posner, M.I. 1967. *Human Performance.* Belmont, CA: Brooks/Cole.

Flanagan, J. 1954. The critical incident technique. *Psychological Bulletin* **51**, 327–358.

Flin, R., O'Connor, P. & Crichton, M. 2008. *Safety at the Sharp End: A Guide to Non-Technical Skills.* Aldershot: Ashgate.

Fogerty, G.J., Saunders, R. & Collyer, R. 1999. Developing a model to predict aircraft maintenance performance. In *Proceedings of the Tenth International Symposium on Aviation Psychology*, R. Jenson (ed.). Columbus, OH: Ohio State University, 1–6.

Gilbert, C., Amalberti, R., Laroche, H. & Paries, J. 2007. Errors and failures: Towards a new safety paradigm. *Journal of Risk Research* **10**(7), 959–975.

Glaser, B.G. & Strauss, A.L. 1967. *The Discovery of Grounded Theory: Strategies for Qualitative Research.* New York: Aldine.

Goggins, R.W., Spielholz, P. & Nothstein, G.L. 2008. Estimating the effectiveness of ergonomics interventions through case studies: Implications for predictive cost-benefit analysis. *Journal of Safety Research* **39**, 339–344.

Gould, S.K., Røed, B.K., Koefoed, V.F., Bridger, R.S. & Moen, B.E. 2006. Performance shaping factors associated with navigation accidents in the Royal Norwegian navy. *Military Psychology* **18**(Suppl), 111–129.

Graeber, R.C. & Marx, D.A. 1993. Reducing human error in aircraft maintenance operations. In *Proceedings of the Flight Safety Foundation International Federation of Airworthiness 46th Annual International Air Safety Seminar*, Arlington, VA: Flight Safety Foundation, 147–160.

Green, R.G., Muir, H., James, M., Gradwell, D. & Green, R. 1996. *Human Factors for Aircrew.* Aldershot: Ashgate.

Grundgeiger, T., Sanderson, P.M. & Dismukes, R.K. 2014. Prospective memory in complex sociotechnical systems. *Zeitschrift für Psychologie* **222**(2), 100–109.

Guynn, M.J., McDaniel, M.A. & Einstein, G.O. 1998. Prospective memory: When reminders fail. *Memory & Cognition* **26**(2), 287–298.

Haque, S. & Conway, M.A. 2001. Sampling the process of autobiographical memory construction. *European Journal of Cognitive Psychology* **13**, 529–547.

Harris, D. & Harris F.J. 2004. Evaluating the transfer of technology between application domains: A critical evaluation of the human component in the system. *Technology in Society* **26**, 551–565.

Hawkins, F.H. 1987. *Human Factors in Flight*. Aldershot: Ashgate, Gower Technical Press.

Health and Safety Executive (HSE). 2016. Health and safety executive statistics, 2015–2016. Available from: http://www.hsfor example,ov.uk/statistics/. (Accessed 21 February 2017).

Health and Safety Executive (HSE). 2017. Health and safety executive cost benefit analysis checklist. Available from: http://www.hsfor example,ov.uk/risk/theory/alarpcheck.htm. (Accessed 3 March 2017).

Heinrich, H.W. 1931. *Industrial Accident Prevention, a Scientific Approach*. New York: McGraw–Hill.

Helmreich, R.L. 2000. On error management: Lessons from aviation. *British Medical Journal* **320**, 745–753.

Helmreich, R.L., Merritt, A.C. & Wilhelm, J.A. 1999. The evolution of crew resource management training in commercial aviation. *International Journal Of Aviation Psychology* **9**(1), 19–32.

Hendrick, H.W. 2003. Determining the cost-benefits of ergonomics projects and factors that lead to their success. *Applied Ergonomics* **34**, 419–427.

Hobbs, A. & Williamson, A. 2003. Associations between errors and contributing factors in aircraft maintenance. *Human Factors* **45**(2), 186–201.

Hollnagel, E. 1993. *Human Reliability Analysis, Context and Control*. London: Academic Press.

Hollnagel, E. 2014. *Safety-I and Safety-II: The Past and Future of Safety Management*. Farnham, UK: Ashgate.

Hollnagel, E., Woods, D.D. & Leveson, N. 2006. *Resilience Engineering: Concepts and Precepts*. Aldershot: Ashgate Publishing.

Hutchins, E. 1995. How a cockpit remembers its speeds. *Cognitive Science* **19**, 265–288.

Hutchins, E. 2001. Distributed cognition. In *The International Encyclopaedia of the Social and Behavioral Sciences*, J.S. Neil & B.B. Paul (eds.). Oxford: Pergamon, 2068–2072.

Johannessen, K.B. & Berntsen, D. 2010. Current concerns in involuntary and voluntary autobiographical memories. *Consciousness & Cognition* **19**, 847–860.

Johnson, W.B. & Avers, K. 2012. Return on investment tool for assessing safety interventions. In *Paper prepared for Shell Aircraft Safety Seminar 2012 Human Factors – Safety's Vital Ingredient*. The Hague, Netherlands.

Joint Service Publication 892 (JSP 892). 2016. Risk management. *Online database (restricted access)*. (Accessed 30 October 2016).

Kanse, L. 2004. Recovery uncovered: How people in the chemical process industry recover from failures. PhD thesis. Eindhoven University of Technology.

Kirwan, B. 1998. Human error identification techniques for risk assessment of high–risk systems – Part 1: Review and evaluation of techniques. *Applied Ergonomics* **29**(3), 157–177.

Klein, G. 2008. Naturalistic decision making. *Human Factors* **50**(3), 456–460.

Kleiner, B.M., Hettinger, L.J., DeJoy, D.M., Huang, Y.-H., & Love, P.E.D. 2015. Sociotechnical attributes of safe and unsafe work systems. *Ergonomics* **58**(4), 635–649.

Kontogiannis, T. 1999. User strategies in recovering from errors in man-machine systems. *Safety Science* **32**, 49–68.

Kontogiannis, T. 2011. A systems perspective of managing error recovery and tactical re-planning of operating teams in safety critical domains. *Journal of Safety Research* **42**(2), 73–85.

Kontogiannis, T. & Malakis, S. 2009. A proactive approach to human error detection and identification in aviation and air traffic control. *Safety Science* **47**(5), 693–706.

Kvavilashvili, L. & Ellis, J. 1996. Varieties of intentions. In *Prospective Memory: Theory and Applications*, M. Brandimonte et al. (eds.). Mahwah, NJ: Lawrence Erlbaum Associates, 23–51.

Kvavilashvili, L. & Mandler, G. 2004. Out of one's mind: A study of involuntary semantic memories. *Cognitive Psychology* **48**, 47–94.

Latorella, K.A. & Prabhu, P.V. 2000. A review of human error in aviation maintenance and inspection. *International Journal of Industrial Ergonomics* **26**(2), 133–161.

Leva, M.C., Kontogiannis, T., Balfe, N., Plot, E. & Demichela, M. 2015. Human factors at the core of total safety management: The need to establish a common operational picture. In *Contemporary Ergonomics and Human Factors 2015. Proceedings of the International Conference on Ergonomics & Human Factors 2015*, S. Sharples et al. (eds.). Daventry, Northamptonshire, UK: Taylor & Francis.

Leveson, N. 2004. A new accident model for engineering safer systems. *Safety Science* **42**, 237–270.

Leveson, N. 2011. *Engineering a Safer World: Systems Thinking Applied to Safety*. Cambridge, MA: MIT Press.

Liang, G.F., Lin, J.T., Hwang, S.L., Wang, E.M.Y. & Patterson, P. 2010. Preventing human errors in aviation maintenance using an on-line maintenance assistance platform. *International Journal of Industrial Ergonomics* **40**(3), 356–367.

Lind, S. 2008. Types and sources of fatal and severe non-fatal accidents. *International Journal of Industrial Ergonomics* **38**, 927–933.

Mace, J.H., Atkinson, E., Moeckel, C.H. & Torres, V. 2011. Accuracy and perspective in involuntary autobiographical memory. *Applied Cognitive Psychology* **25**, 20–28.

Malakis, S., Kontogiannis, T. & Kirwan, B. 2010. Managing emergencies and abnormal situations in air traffic control (part I): Taskwork strategies. *Applied Ergonomics* **41**(4), 620–627.

Mandler, G. 1985. *Cognitive Psychology: An Essay in Cognitive Science*. Hillsdale, NJ: Erlbaum.

Marsh, R.L., Hicks, J.L. & Bink, M.L. 1998. Activation of completed, uncompleted, and partially completed intentions. *Journal of Experimental Psychology: Learning, Memory & Cognition* **24**, 350–361.

Matsika, E., Ricci, S., Mortimer, P., Georgiev, N. & O'Neill, C. 2013. Rail vehicles, environment, safety and security. *Research in Transportation Economics* **41**, 43–58.

Matthews, B.W. 1975. Comparison of the predicted and observed secondary structure of T4 phage lysozyme. *Biochimica et Biophysica Acta (BBA) – Protein Structure* **405**(2), 442–451.

Maurino, D.E., Reason, J., Johnston, N. & Lee, R.B. 1995. *Beyond Aviation Human Factors: Safety in High Technology Systems*. Aldershot: Avebury Aviation.

Mazzoni, G., Vannucci, M. & Batool, I. 2014. Manipulating cues in involuntary autobiographical memory: Verbal cues are more effective than pictorial cues. *Memory and Cognition* **42**, 1076–1085.

Mecacci, L. & Righi, S. 2006. Cognitive failures, metacognitive beliefs and aging. *Personality and Individual Differences* **40**, 1453–1459.

Micheli, G. & Cagno, E. 2009. Perception of safety issues and investments in safety management in small and medium-sized enterprises: A survey in the Lecco area. *Prevention Today* **4**(10), 7–18.

Morel, G., Amalberti, R. & Chauvin, C. 2008. Articulating the differences between safety and resilience: The decision-making process of professional sea-fishing skippers. *Human Factors* **50**(1), 1–16.

Mullan, B., Smith, L., Sainsbury, K., Allom, V., Paterson, H. & Lopez, A.-L. 2015. Active behaviour change safety interventions in the construction industry: A systemic view. *Safety Science* **79**, 139–148.

Murphy, L.A., Robertson, M.M. & Carayon, P. 2014. The next generation of macroergonomics: Integrating safety climate. *Accident Analysis & Prevention* **68**, 16–24.

Naderpour, M., Lu, J. & Zhang, G. 2014. A situation risk awareness approach for process systems safety. *Safety Science* 64, 173–189.

Neisser, U. 1976. *Cognition and Reality: Principles and Implications of Cognitive Psychology*. San Francisco, CA: Freeman.

Nikolic, M.I. & Sarter, N.B. 2007. Flight deck disturbance management: A simulator study of diagnosis and recovery from breakdowns in pilot-automation coordination. *Human Factors* 49(4), 553–563.

Niskanen, T., Louhelainen, K. & Hirvonen, M.L. 2016. A systems thinking approach of occupational safety and health applied in the micro-, meso- and macro-levels: A Finnish survey. *Safety Science* **82**, 212–227.

Norman, D. & Bobrow, D. 1975. On data-limited and resource-limited processes. *Cognitive Psychology* **7**, 44–64.

Norman, D. & Shallice, T. 1986. Attention to action: Willed and automatic control of behavior. In *Consciousness and Self-Regulation: Advances in Research.*, R. Davidson et al. (eds.). Plenum: New York.

Norman, D.A. 1981. Categorization of action slips. *Psychological Review* **88**, 1–15.

Norman, D.A. 1993. *Things that Make Us Smart: Defending Human Attributes in the Age of the Machine*. Reading, MA: Addison-Wesley.

Oppenheim, A.N. 1992. *Questionnaire Design, Interviewing and Attitude Measurement*. London: Pinter.

Palmer, H.M. & McDonald, S. 2000. The role of frontal and temporal lobe processes in prospective remembering. *Brain and Cognition* **44**, 103–107.

Patankar, M.S. & Taylor, J.C. 2004. *Applied Human Factors in Aviation Maintenance*. Aldershot, UK: Ashgate Publishers.

Patel, V.L., Cohen, T., Murarka, T., Olsen, J., Kagita, S., Myneni, S., Buchman, T. & Ghaemmaghami, V. 2011. Recovery at the edge of error: Debunking the myth of the infallible expert. *Journal of Biomedical Informatics* **44**(3), 413–424.

Perrow, C. 1999. *Normal Accidents: Living with High-Risk Technologies*. Princeton, NJ: Princeton University Press, updated edition.

Plant, K.L. & Stanton, N.A. 2013a. What is on your mind? Using the perceptual cycle model and critical decision method to understand the decision-making process in the cockpit. *Ergonomics* **56**(8), 1232–1250.

Plant, K.L. & Stanton, N.A. 2013b. The explanatory power of schema theory: Theoretical foundations and future applications in ergonomics. *Ergonomics* **56**(1), 1–15.

Plant, K.L. & Stanton, N.A. 2012. Why did the pilots shut down the wrong engine? Explaining errors in context using schema theory and the perceptual cycle model. *Safety Science* **50**, 300–315.

Plant, K.L. & Stanton, N.A. 2016. *Distributed Cognition and Reality: How Pilots and Crews Make Decisions*. Boca Raton, FL: CRC Press.

Polit, D.F. & Beck, C.T. 2004. *Nursing Research: Principles & Methods*. Philadelphia, PA: Lippincott, Williams & Wilkins, seventh edition.

Prabhu, P. & Drury, C.G. 1992. A framework for the design of the aircraft inspection information environment. In *Proceedings of the 7th FAA Meeting on Human Factors Issues in Aircraft Maintenance and Inspection*, Atlanta, GA: FAA, Office of Aviation Medicine, 54–60.

Rafferty, L.A., Stanton, N.A., & Walker, G.H. 2013. Great expectations: A thematic analysis of situation in fratricide. *Safety Science* **56**, 63–71.

Rankin, B. & Allen, J. 1996. Boeing introduces MEDA, maintenance error decision aid. *Airliner* April–June, 20–27.

Rashid, H.S.J., Place, C.S. & Braithwaite, G.R. 2010. Helicopter maintenance error analysis: Beyond the third order of the HFACS-ME. *International Journal of Industrial Ergonomics* **40**(6), 636–647.

Rasmussen, A.S. & Berntsen, D. 2011. The unpredictable past: Spontaneous autobiographical memories outnumber autobiographical memories retrieved strategically. *Consciousness & Cognition* **20**, 1842–1846.

Rasmussen, J. 1982. Human errors: A taxonomy for describing human malfunctions in industrial installations. *Journal of Occupational Accidents* **4**, 311–333.

Rasmussen, J. 1997. Risk management in a dynamic society: A modelling problem. *Safety Science* **27**(2), 183–213.

Rasmussen, J. & Pedersen, O.M. 1984. Human factors in probabilistic risk analysis and risk management. In *Human Error*, J. Reason (ed.). 1990. Cambridge: Cambridge University Press.

Reason, J. 1990. *Human Error.* Cambridge: Cambridge University Press.

Reason, J. 1997. *Managing the Risks of Organizational Accidents.* Aldershot: Ashgate.

Reason, J. 2008. *The Human Contribution: Unsafe Acts, Accidents and Heroic Recoveries.* Aldershot: Ashgate.

Reason, J. & Hobbs, A. 2003. *Managing Maintenance Error: A Practical Guide.* Aldershot: Ashgate.

Reiman, R. 2011. Understanding maintenance work in safety-critical organisations – managing the performance variability. *Theoretical Issues In Ergonomics Science* **12**(4), 339–366.

Robson, C. 2011. *Real World Research.* Chichester: Wiley, third edition.

Rowntree, D. 1981. *Statistics Without Tears.* London: Penguin.

Rumelhart, D.E. & Norman, D.A. 1983. Representation in Memory. *CHIP Technical Report 116.* San Diego, CA: Center for Human Information Processing, University of California. In *Memory*, G. Cohen et al. (eds.). 1986. Milton Keynes: Open University Press.

Salmon, P.M., Walker, G.H. & Stanton, N.A. 2016. Pilot error versus sociotechnical systems failure: A distributed situation awareness analysis of Air France 447. *Theoretical Issues in Ergonomics Science* **17**(1), 64–79.

Sarter, N.B. & Alexander, H.M. 2000. Error types and related error detection mechanisms in the aviation domain: An analysis of aviation safety reporting system incident reports. *International Journal of Aviation Psychology* **10**, 189–206.

Saward, J.R.E. & Jarvis, S. 2007. *Causal Factors Influencing UK Military Aviation Incidents.* School of Engineering: Cranfield University.

Saward, J.R.E. & Stanton, N.A. 2015a. Individual latent error detection and recovery in naval aircraft maintenance: Introducing a proposal linking schema theory with a multi-process approach to human error research. *Theoretical Issues in Ergonomics Science* **16**(3), 255–272.

Saward, J.R.E. & Stanton, N.A. 2015b. Individual latent error detection: Is there a time and a place for the recall of past errors? *Theoretical Issues in Ergonomics Science* **16**(5), 533–552.

Saward, J.R.E. & Stanton, N.A. 2017. Individual latent error detection: A golden two hours for detection. *Applied Ergonomics* **59**, 104–113.

Schluter, J., Seaton, P. & Chaboyer, W. 2008. Critical incident technique: A user's guide for nurse researchers. *Journal of Advanced Nursing* **61**(1), 107–114.

Sellen, A.J. 1994. Detection of everyday errors. *Applied Psychology: An International Review* **43**, 475–498.

Sellen, A.J., Louie, G., Harris, J.E. & Wilkins, A.J. 1996. What brings intentions to mind? An *in situ* study of prospective memory. *Memory* **5**, 483–507.

Shappell, S.A., & Wiegmann, D.A. 2009. A methodology for assessing safety program stargeting human error in aviation. *The International Journal of Aviation Psychology*, **7**(1), 252–269.

Shorrock, S.T. & Kirwan, B. 2002. Development and application of a human error identification tool for air traffic control. *Applied Ergonomics* **33**, 319–336.

Smallwood, J. & Schooler, J.W. 2006. The restless mind. *Psychological Bulletin* **132**, 946–958.

Smith, K. & Hancock, P.A. 1995. Situation awareness is adaptive, externally directed consciousness. *Human Factors* **37**, 137–148.

Smith, R.E. 2003. The cost of remembering to remember in event based prospective memory: Investigating the capacity demands of delayed intention performance. *Journal of Experimental Psychology: Learning, Memory, & Cognition* **29**, 347–361.

Smith, R.E., Hunt, R.R., McVay, J.C. & McConnell, M.D. 2007. The cost of event-based prospective memory: Salient target events. *Journal of Experimental Psychology: Learning, Memory and Cognition* **33**, 734–746.

Stanton, N. & Baber, C. 1996. A systems approach to human error identification. *Safety Science* **22**(1–5), 215–228.

Stanton, N.A. & Baber, C. 2003. On the cost-effectiveness of ergonomics. *Applied Ergonomics* **34**, 407–411.

Stanton, N.A. & Harvey, C. 2017. Beyond human error taxonomies in assessment of risk in sociotechnical systems: A new paradigm with the EAST 'broken-links' approach. *Ergonomics* **60**(2), 221–233.

Stanton, N.A. & Salmon, P.M. 2009. Human error taxonomies applied to driving: A generic driver error taxonomy and its implications for intelligent transport systems. *Safety Science* **47**(2), 227–237.

Stanton, N.A., Salmon, P., Harris, D., Marshall, A., Demagalski, J., Young, M.S., Waldmann, T. & Dekker, S. 2009a. Predicting pilot error: Testing a new methodology and a multi-methods and analysts approach. *Applied Ergonomics* **40**(3), 464–471.

Stanton, N.A., Salmon, P.M. & Walker, G.H. 2015. Let the reader decide: A paradigm shift for situation awareness in sociotechnical systems. *Journal of Cognitive Engineering and Decision Making* **9**(1), 44–50.

Stanton, N.A., Salmon, P.M., Walker, G.H. & Jenkins, D. 2009b. Genotype and phenotype schemata and their role in distributed awareness in collaborative systems. *Theoretical Issues in Ergonomics Science* **10**(1), 43–68.

Stanton, N.A. & Walker, G.H. 2011. Exploring the psychological factors involved in the Ladbroke Grove rail accident. *Accident Analysis and Prevention* **43**, 1117–1127.

Stanton, N.A. & Young, M.S. 1999. What price ergonomics? *Nature* **399**, 197–198.

Sunderland, A., Harris, J. & Baddeley, A. 1983. Do laboratory tests predict everyday memory? A neuropsychological study. *Journal of Verbal Learning and Verbal Behavior* **22**, 341–357.

Thomas, M.J.W. 2004. Predictors of threat and error management: Identification of core non-technical skills and implications for training systems design. *International Journal of Aviation Psychology* **14**, 207–231.

Trafford, V. & Leshem, S. 2008. *Stepping Stones to Achieving Your Doctorate: By Focusing on Your Viva from the Start: Focusing on Your Viva from the Start*. UK: McGraw-Hill Education.

Tulving, E. 1983. *Elements of Episodic Memory*. New York: Oxford University Press.

Twelker, P.A. 2003. The critical incident technique: A manual for its planning and implementation. *Journal of Advanced Nursing* **61**(1), 107–114.

Van den Berg, S.M., Aarts, H., Midden, C. & Verplanken, B. 2004. The role of executive processes in prospective memory tasks. *European Journal of Psychology* **16**(4), 511–533.

Vaughan, D. 1997. *The Challenger Launch Decision*. Chicago: University of Chicago Press.

Walker, G.H., Stanton, N.A., Salmon, P.M. & Jenkins, D.P. 2008. A review of sociotechnical systems theory: A classic concept for new command and control paradigms. *Theoretical Issues in Ergonomics Science* **9**(6), 479–499.

Wallace, J.C., Kass, S.J. & Stanny, C.J. 2002. The cognitive failures questionnaire revisited: Dimensions and correlates. *The Journal of General Psychology* **129**(3), 238–256.

Weick, K. 1987. Organizational culture as a source of high reliability. *California Management Review* **29**, 112–127.

Wiegmann, D.A. & Shappell, S.A. 2001. Human error perspectives in aviation. *International Journal of Aviation Psychology* **11**(4), 341–357.

Wiegmann, D.A. & Shappell, S.A. 2003. *A Human Error Approach to Aviation Accident Analysis: The Human Factors Analysis and Classification System*. Ashgate: Aldershot.

Wilkinson, W.E., Cauble, L.A. & Patel, V.L. 2011. Error detection and recovery in dialysis nursing. *Journal of Patient Safety* **7**(4), 213–223.

Wilson, J.R. 2014. Fundamentals of systems ergonomics/human factors. *Applied Ergonomics* **45**(1), 5–13.

Wioland, L. & Amalberti, R. 1996. When errors serve safety: Towards a model of ecological safety, CSEPC 96. *Cognitive Systems Engineering in Process Control*. Kyoto, Japan, 184–191.

Woo, D.M. and Vicente, K.J. 2003. Sociotechnical systems, risk management, and public health: Comparing the North Battleford and Walkerton outbreaks. *Reliability Engineering and System Safety* **80**, 253–269.

Woods, D.D. 1984. Some results on operator performance in emergency events. In *Ergonomic Problems in Process Operations. Institute of Chemical Engineering Symposium*, D. Whitfield (ed.). **90**, 21–31.

Woods, D.D., Dekker, S., Cook, R., Johannesen, L. & Sarter, N. 2010. *Behind Human Error*. Farnham: Ashgate second edition.

Woods, D.D. & Hollnagel, E. 2006. *Joint Cognitive Systems: Patterns in Cognitive Systems Engineering*. London: Taylor & Francis.

Zapf, D. & Reason, J.T. 1994. Introduction: Human errors and error handling. *Applied Psychology: An International Review* **43**, 427–432.

Zink, K.J., Hendrick, H. & Kleiner, B. 2002. A vision of the future of macro-ergonomics. In *Macroergonomics: Theory, Methods, and Applications*, W.H. Hal, K. Brian (eds.), 347–358.

Index